– GREG APPEL | EDDY JOKOVICH | RUSTY CODEC –

VISITOR'S GUIDE TO THE MOON

GUTHUGGA PIPELINE PRESS

Visitor's Guide To The Moon
ISBN (paperback):   978-1-7635701-1-5
ISBN (Amazon):   979-8-3268763-3-1

©2024 Greg Appel & Eddy Jokovich

All rights reserved. No part of this book may be reproduced in any form or by any electronic or mechanical means, including information storage and retrieval systems, without written permission from the authors, except for the use of brief quotations in book reviews and promotional material.

June 2024. Published by Guthugga Pipeline Press.

Guthugga Pipeline Press
PO Box 1265, Darlinghurst NSW 1300

Production: ARMEDIA
Assorted lunar photography provided by Blake Compton. Archival and assorted space travel photographs provided by NASA.

Published and produced on the lands of the Wangal and Gadigal people.

EDITORIAL NOTE ON THE USE OF AI TECHNOLOGY
We employ artificial intelligence tools in the editing process of our articles. These tools assisted with transcriptions of audio recordings, grammar correction, refinement and formatting.

A catalogue record for this work is available from the National Library of Australia

Contents

1. Blast off, eat, sleep, leap: The story of the first moon landing............9
The Cold War...............10
Why go to the moon?...............11
It's party time...............12
Star warriors: The Apollo 11 crew...............15
The journey...............17
The step...............19
Getting back...............22
Life after walking on the moon...............23

2. But did the moon landing really happen?...............26
Historical context...............29
Scepticism and conspiracy theories...............30
Anomalies in photographs and videos...............32
Technological capabilities...............34
Political motivations...............37
Independent verification...............39
Scientific analysis of moon rocks...............40
Testimonies from astronauts and space scientists...............42
Technological evidence...............43
Global consensus...............44
Why do the conspiracies continue?...............45

3. They're back! A return to the moon...............48
The coalition for lunar exploration...............50
The pioneers: roving the lunar wastelands...............52
The Lunar Communication Link...............53
The moon base...............54
Getting the kids early...............55
Finally: here come the humans... again...............56
Looking forward...............58

4. The astronomical and geological history of the moon ... 60
Formation ... *63*
The early state of the moon ... *65*
Cratered appearance ... *66*
Maria and highlands ... *68*
Moon dust ... *72*
Volcanic activity ... *74*
Tectonic shifts ... *76*
Apollo missions ... *77*
How old is the solar system? ... *78*

5. Illuminating the divine: Moon worship across cultures ... 81
The luminous god: Early human cultures and lunar reverence ... *84*
The cradle of civilisation: Moon worship in the Ancient Near East ... *86*
The moon in Ancient Egypt ... *88*
The moon becomes female: Greco–Roman mythology ... *89*
The Eastern luminary: Lunar deities of Asia ... *93*
The moon's reflection in Indigenous traditions ... *94*
The influence in the transition of beliefs: polytheism to monotheism ... *95*

6. Moon mythology, werewolves, vampires and romantics ... 98
The moon as a mythological symbol ... *101*
Association with the underworld ... *103*
Connection to mythical creatures ... *104*
Werewolf v. vampire ... *106*
Romantics, Gothics and the moon ... *109*
Fate ... *112*
The moon as a symbol of fascination and fear ... *112*

7. Howling: a moonlit madness unravelled ... 115
A historical perspective: a short history of lunacy and lore ... 117
Scientific inquiry: Separating fact from folklore ... 119
Psychological perspectives: Moonlit reflections ... 123
Scepticism and critique: A rational look at lunar effects ... 126
Personal stories: Moonlit testimonies ... 128
Modern moon howlers ... 131

8. The relationship between the moon and the tides ... 133
The weird force of gravity ... 136
It gets weirder: gravitons ... 138
The gravitational influence of the moon ... 139
Spring tides ... 141
The impact of tides on marine ecosystems ... 143
The influence on coastal regions ... 144
How the tides affect navigation and influenced human activity ... 146
How understanding the moon and tides has changed humanity ... 147
Tides and creatures ... 148
Money making tides ... 149
How does our moon's gravitational pull stack up with others? ... 152

9. The new space race ... 155
The moon as a strategic outpost ... 156
Military and strategic interests ... 159
Want to buy a block on the moon while you still can? ... 160
The economics of lunar exploration: making real earth money ... 162
International collaborations and tensions ... 164
International space pals ... 165
Who owns the moon? ... 166

10. The mega rich in space 169
A new era of space exploration *170*
The prominent figures of the space exploration *174*
Are these the right people to be involved in space exploration? 176
Beyond the orbit of the Earth *179*
The shift from government to private sector *180*
Space tourism *182*
Space colonisation on the moon and other planets *184*
The future and development of the space economy *188*
Policy and regulation implications for moon and space exploration *189*
Technological innovation *191*

11. A holiday on the moon: is it really possible? 194
Lunar holiday *196*
Lunar tourist attractions *198*
Accommodation and facilities *199*
Travel to and from the moon *200*
Is the moon a healthy environment for humans? *201*
How much is this going to cost? *203*
Legal and technological obstacles *204*
Realistic or not? *205*

12. Glam rock and a cultural loop around the moon 207
What was glam rock? *212*
Glam films *215*
More moon films *217*
Glam fashion: Moon boots *219*
Moon songs *221*
The symbolism of the moon in music *221*
Cultural perspectives *223*
The return of the glam *227*

13. The last man on the moon and the first woman…..........228

The last man on the moon.. 231

The Artemis program... 233

NASA and friends.. 235

The first woman on the moon 237

Artemis v. Apollo..238

The others 239

14. Who's going to the moon? Machines v. humans.............................242

An imaginary international moon meeting..................................... 244

CHAPTER 1

Blast off, eat, sleep, leap: The story of the first moon landing

It's hard to imagine getting a little sleep on your way to the first moon landing. But that's just what the three astronauts did after they were blasted into space and compressed by extreme G forces on July 16, 1969.

> *July 16 20:33:34 hrs, Mission Control Center, Houston:*
> *Roger, 11. That was a good readback...We'd like to just say that on a flight plan update here, just to remind you of some things, and you can do them at your convenience and then go to sleep early if you'd like. We don't have anything else planned, but we'd like to just remind you on the filter change, the O2 fuel cell purge. And we'd like a LM/CM, DELTA-P... (and so on... NASA loves acronyms!)*

And off to sleep they went. The first lunar landing consisted of moments of terror, humour, wonder and long stretches of boredom. It was an amazing technical achievement and an absolute OH&S nightmare. As we all know, except for those of a conspiratorial bent, they made it. How it all happened is a very positive story that came out of a dark time for humanity. Instead of raining nuclear bombs on each other, we sent bombs with humans on top of them—all the way to the moon. And somehow managed to get them back.

The Cold War

Above the mists of the Cold War, a battle took place. This was a race not just of nations, but of ideologies, with the United States and the Soviet Union locked in a fierce competition to assert dominance beyond this planet. Amid this tense geopolitical chess game, NASA's Apollo program emerged, not merely as a scientific endeavour, but as a crucial pawn in the power struggle between these two superpowers. *Apollo v Soyuz*. In particular, the *Apollo 11* mission bore the weight of this global rivalry, its objectives entwined with the desire to claim the ultimate high ground—the moon—and put Americans on it.

As preparations for the mission unfolded, the world watched, unaware of the complex web of espionage, technological sabotage, and political manoeuvring that underpinned the journey to the moon. The *Apollo 11* crew—Neil Armstrong, Buzz Aldrin and Michael Collins—were not just astronauts; they were the chosen few who would carry the burden of America's national pride and the unspoken fears of a conflict that threatened to escalate beyond the confines of Earth. The mission's objectives were clear. Yet, beneath these goals lay the immense pressure to succeed at all costs, to plant the American flag on the lunar surface before the Soviets could claim victory in the space race.

This backdrop of intrigue cast a strange shadow over the *Apollo 11* mission. It was a time when every satellite launch was scrutinised for its military potential, and every broadcast from space was a volley in the propaganda war. The Russians had clearly beaten the Americans to both sending satellites and humans into orbit around the earth, with *Sputnik 1* and *Vostok 1*. Yuri Gagarin was the first human in space, after they'd sent a succession of unfortunate dogs and monkeys out there. They seemed to be about to land a cosmonaut on the moon. The space race was as much about demonstrating technological superiority and the capability to deploy ballistic missiles as it was about exploring new frontiers. In this charged atmosphere, the *Apollo 11* mission became a beacon of hope for some, and for others, the weaponisation of space. This was the real star wars.

Why go to the moon?
On the face of it, the *Apollo 11* mission aimed to achieve President John F. Kennedy's goal set on May 25, 1961: to land humans on the moon and return them to Earth. But there were big hurdles to overcome to get there.

The Apollo moon missions, particularly *Apollo 11*, were subjects of controversy regarding their immense financial cost. Critics argued that the billions of dollars spent could have been directed towards solving pressing societal issues on Earth, such as poverty, hunger, and healthcare. The debate was fueled by the Cold War context, with some viewing the space race as an extravagant display of technological prowess rather than a pursuit of scientific knowledge.

This contention reflected broader questions about government spending priorities and the allocation of resources. While proponents of the program highlighted its technological advancements, contributions to science, and national pride, sceptics remained concerned about the opportunity costs. The discourse around the expense of sending men to the moon underscores an ongoing conversation about the balance between exploration and addressing immediate challenges on Earth.

The financial journey of the Apollo moon missions, culminating in the historic *Apollo 11* landing, was as monumental as the technical challenges overcome by NASA. The Apollo program's total cost was approximately \$US25.4 billion—equivalent to over \$150 billion in today's dollars—a figure that underscores the colossal investment in taking humanity to the moon and back. This funding was approved by the U.S. Congress in a series of appropriations throughout the 1960s—the Cold War climate significantly influenced Congressional approval, as lawmakers were swayed by the argument that superiority in space would translate to military and ideological advantages.

The program's funding was also a testament to President Kennedy's persuasive vision of American leadership in space, science, and technology. Despite debates over domestic spending, the Apollo budget passed through Congress, reflecting a strong bipartisan consensus on the strategic importance of space exploration and the national prestige associated with landing the first humans on the moon. J.F.K. would not live to see it, but somehow the mission was go.

It's party time

As the *Apollo 11* mission prepared to launch in July 1969, a palpable sense of excitement and anticipation swept across not just the United States, but the entire world. Near the Kennedy Space Center in Florida, vast crowds gathered, drawn by the promise of witnessing a monumental moment in human history. The crowd was estimated at 1 million, causing some extreme pressures on the local tourist infrastructure. People from all walks of life—families, space enthusiasts, journalists, and international visitors—converged on beaches, roadsides, and any vantage points they could find, eager to be part of the collective experience. The atmosphere was electric, a mixture of nervous anticipation and jubilant expectation, as the countdown to launch echoed across the gathered throng.

While the astronauts were isolated from any humans in preparation for the flight—there was a drunken orgy going on around them. Around the world, bars and restaurants seized the moment, crafting themed cocktails that captured the essence of the mission. Drinks like the "Moonwalk" cocktail, invented by bartender Joe Gilmore at the Savoy Hotel in London:

1 part grapefruit juice
1 part orange curaçao
1 dash rose water
2 parts sparkling white wine or sparkling rosé wine

Instructions:
If you're using sparkling wine, shake all the other ingredients with ice first. Strain into a flute glass and top with sparkling wine. If you're using vodka, shake all the ingredients with ice and strain into a chilled cocktail glass.

These celebratory concoctions allowed people to toast the astronauts' journey, blending the excitement of space exploration with the conviviality of shared human experiences. It was like a huge fireworks night! The negatives were all washed away with the cocktails. The launch of *Apollo 11* turned into an international celebration, a moment when time seemed to stand still as millions watched on television or listened on the radio, their breaths held as the giant *Saturn V* rocket ignited. So tall that the weather at the top was considerably different to

1. Michael Collins. 2. Neil Armstrong, Michael Collins, 'Buzz' Aldrin. 3. Neil Armstrong. 4. Laika, the first dog in space. 5. Watching the takeoff, Kennedy Space Center, Florida. 6. 'Buzz' Aldrin. 7. President John F. Kennedy inspecting a moon capsule. 8. "Moonwalk" cocktail. 9. View of Earth from moon, 1969.

below. Over 110 metres! The collective joy and pride felt during the successful launch was a high point for the era. Not just in America—but around the world there were viewing parties, and the sense of unity and hope was palpable.

For those gathered near the launch site, the experience was sensory overload—the ground shaking, the roar of the engines, and the sight of the rocket ascending into the sky, a bright beacon of human ambition and ingenuity. The *Apollo 11* launch was not just a scientific milestone; it was a cultural event that captured the imagination of the world, a shared moment of achievement and possibility. But there was still a lot that could go wrong.

Star warriors: The Apollo 11 crew

In the transcripts of the *Apollo 11* voyage, the crew comes across as a bunch of tech guys who happen to be going to the moon.

> ***17 July 18:10:18 hrs, CMP (Command Module Pilot):*** *That's why I've been eating so much today. I haven't had anything to do. He won't let me touch it any more!*
> ***CC (Control Center):*** *Roger.*
> ***CC (Control Center):*** *11, Houston. If that's not the Earth, we're in trouble.*
> ***CDR (Commander):*** *That's the Earth, and we have a very good view of it today. There are a few more cloud bands on than yesterday when we beamed down to you, but it's a beautiful sight.*

The heroes of the moon mission are interesting characters. At the time, they felt they were part of a larger Apollo team and the whole thing had a quasi military feel to it. Being selected for this particular mission thrust them in front of an international audience—they became more than mortal, and sometimes found it hard back on earth as living gods.

Neil Armstrong: Born on August 5, 1930, in Wapakoneta, Ohio. Armstrong would make history as the first person to walk on the moon. Before joining NASA, Armstrong served as a naval aviator from 1949 to 1952 and participated in the Korean War. He joined NASA's astronaut corps in 1962 and was command pilot for his first mission, *Gemini 8*, in 1966, where he performed the first successful docking of two vehicles in space. Armstrong's calm demeanour and precise engineering skills made him the ideal candidate to command *Apollo 11*, the first manned lunar landing mission. He became a living American hero, a role he would grow to hate.

'Buzz' Aldrin: Born Edwin Eugene Aldrin Jr. on January 20, 1930, in Montclair, New Jersey, Aldrin was the lunar module pilot on *Apollo 11* and the second person to walk on the moon. His odd nickname has become iconic, inspiring the animated character Buzz Lightyear. It came from his baby sister trying to say 'brother'—and coming out with 'buzzar'. It was like Buzz was meant to be an astronaut. Buzz's mother was Marion Moon, before she married Edwin Aldrin, a pioneer aviator. Before his astronaut career, Aldrin was a fighter pilot in the U.S. Air Force and served in the Korean War.

He earned a Doctorate of Science in Astronautics at MIT, writing his thesis on manned orbital rendezvous, a topic that proved crucial to the success of the Apollo missions. Aldrin was selected by NASA in 1963 and first flew on *Gemini 12*, the final mission of the Gemini program, where he set a new record for extravehicular activity. Tragically, a year before he walked on the moon, Buzz's mother committed suicide. Aldrin was deeply religious and took his religion on to the moon, taking communion at the same time it was happening in his church on Earth. Buzz became an international hero after the landing—but it didn't sit comfortably with him: "The difficult part of my life was not going to the moon; it was what I had to face when I returned".

Michael Collins: Born on October 31, 1930, in Rome, Italy. Collins served as the command module pilot for *Apollo 11*. Before joining NASA in 1963, Collins was an experimental test pilot and a major in the U.S. Air Force. He first flew in space as the pilot of *Gemini 10*, during which he performed two spacewalks. Unlike his crewmates, Collins did not walk on the moon, and obligingly took the job that would not really go down in history—waiting for the guys to walk on the moon and come back to the command module *Columbia* as it circled. Nevertheless, driving the lunar bus was a critical role.

> Collins (from his memoirs): "I am alone now, truly alone, and absolutely isolated from any known life. I am it. If a count were taken, the score would be three billion plus two over on the other side of the moon, and one plus God knows what on this side."

The journey

This voyage was not only a trajectory through space; it was a meticulously planned series of manoeuvres, an intersection of celestial mechanics and human ingenuity. The enormous *Saturn V* rocket, the most powerful machine ever built by humankind, capable of breaking the gravitational shackles of Earth. Once in Earth's orbit, the spacecraft executed the trans-lunar injection, a critical burn of the third-stage engine that set *Apollo 11* on its course for the moon. This manoeuvre required precise calculations and timing; any deviation could jeopardise the mission's success.

As the Earth receded into the distance, becoming nothing more than a vibrant blue and green marble, the crew experienced the profound isolation and beauty of space. So the Earth below could see all of this, they also moonlighted as a camera team.

> *July 16 20:09:36 hrs, CC (Control Center): Hello, Apollo 11. Houston. We'd like you to keep the TV on for about 10 minutes or so, so we can get some good comparison on the camera. You can do anything your heart desires on the TV: interior, exterior, pan in and out, anything you'd like. Over.*
> *CC (Control Center): 11, Houston. Over.*

The images from these Apollo missions inspired many people on Earth. One positive spinoff was the birth of the environmental movement. Yes, it was a costly mission, but this money didn't go nowhere—It employed myriads of people and washed through various economies, especially swampy Florida.

The *Apollo 11* spacecraft was a marvel of engineering, comprising two distinct components: the *Columbia* command module and the *Eagle* lunar module. The *Columbia*, piloted by Michael Collins, served as the crew's quarters and control centre for the duration of the mission. Its conical design was optimised for re-entry into Earth's atmosphere, hopefully ensuring the safety of the crew upon their return. Outfitted with navigation, communication, and life support systems, the *Columbia* was the astronauts' lifeline, a sanctuary in the void of space.

The *Eagle* lunar module, on the other hand, was the vessel that would make history. A technological wonder, it was designed specifically for the lunar descent and ascent, a testament to human engineering and the dream of walking on the moon. Its spidery appearance, with thin landing legs and a boxy structure, was a stark contrast to the sleek aerodynamics of Earth-bound vehicles. Armstrong and Aldrin piloted the *Eagle*, detaching from the *Columbia* to begin their descent to the lunar surface. There was a lot of debate about naming these crafts but 'eagle' was a pretty obvious choice for an all-American team.

As the lunar module touched down in the Sea of Tranquility, the words "the Eagle has landed" echoed across the globe, marking the culmination of a journey that had captured the hopes and dreams of humanity.

The step

The world watched in wonder and anticipation as Neil Armstrong descended the *Eagle*'s ladder to set foot on the lunar surface, marking the moment with words that would resonate through history:

> *"That's one small step for [a] man, one giant leap for mankind."*

The 'a' in brackets became the cause of much debate over the years. Armstrong himself stated that he believed he had said "a man," suggesting that the transmission limitations or the stress of the moment might have obscured the word. The debate over this single letter has persisted, with analyses and digital enhancements of the audio recording attempting to resolve the issue. Anyone who listened to the original crackly transmission will know that it's lost in the moon dust somewhere.

The lunar surface activities of Armstrong and Aldrin were a blend of scientific inquiry and symbolic gestures. The astronauts set up an array of experiments designed to provide insights into the moon's geology, seismic activity, and solar wind. The deployment of the lunar laser ranging retroreflector, for instance, allowed scientists back on Earth to measure the distance to the moon with unprecedented accuracy, providing valuable data for decades to come.

Amidst these scientific endeavours, the planting of the American flag is an enduring image. A very big moment for American imperialism, possibly the zenith of its empire. The flag planting, coupled with the collection of lunar rock and soil samples, underscored the dual nature of the *Apollo 11* mission. Science and hard diplomacy.

Then there were the strange moments, such as the missing 10 minutes when Armstrong was on the moon but 'off coms'. Though there is much speculation, no one really knows what he was up to.

America had a certain President Richard Nixon at the helm during the *Apollo 11* mission and he made a call while they moonwalked.

> **20 July 23:48:25 hrs, CC (Control Center):** *Go ahead, Mr. President. This is Houston. Out.*
> **President Nixon:** *Neil and Buzz, I am talking to you by telephone from the Oval Room at the White House, and this certainly has to be the most historic telephone*** [three asterisks denote clipping of words and phrases] call ever made. I just can't tell you how proud we all are of what you *** for every American. This has to be the proudest day of our lives. And for people all over the world, I am sure they, too, join with Americans in recognising what an immense feat this is. Because of what you have done, the heavens have become a part of man's world. And as you talk to us from the Sea of Tranquility, it inspires us to redouble our efforts to bring peace and tranquility to Earth. For one priceless moment in the whole history of man, all the people on this Earth are truly one; one in their pride in what you have done, and one in our prayers that you will return safely to Earth.*

1. The first moon landing. 2. The Apollo 11 takeoff. 3. Mission Control, NASA. 4. The Columbia splashdown in the Pacific Ocean. 5. Lunar module ascent stage, with Earth in the background. 6. Unidentified space object: What did 'Buzz' Aldrin really see out there? 7. President Richard Nixon at the Oval Office, speaking to Neil Armstrong through the NASA Control Center.

CDR (EVA): *Thank you, Mr. President. It's a great honour and privilege for us to be here representing not only the United States but men of peace of all nations, and with interest and a curiosity and a vision for the future. It's an honour for us to be able to participate here today.*

Getting back

After their internationally televised activities on the lunar surface, Armstrong and Aldrin rejoined Collins, who had been orbiting the moon in the *Columbia* command module. The ascent from the moon began with the ignition of the *Eagle's* ascent engine, a critical act that propelled them out of the lunar gravity field and into lunar orbit. The rendezvous and docking with *Columbia* went very well. This moment marked the reunion of the *Apollo 11* crew, now ready to embark on their journey back to Earth, carrying with them precious lunar samples and invaluable data.

President Nixon organised a world tour, to bask in the love from this mission, and made sure he would be waiting in the right place when the mission (hopefully) splashed down.

The journey back to Earth was a mirror reflection of their outbound voyage, filled with some boredom, and some anxious anticipation. Now with the added weight of their successful mission. On July 24, 1969, the world watched with bated breath as the *Columbia* command module re-entered Earth's atmosphere, its heat shield withstanding the intense heat of re-entry. The splashdown in the Pacific Ocean, was a moment of jubilation, a symbol of the mission's successful conclusion. The *USS Hornet*, the primary recovery ship, was waiting nearby.

It was a strange re-entry into earthly life. Navy divers were the first people to greet the astronauts: "[we] brought the contamination suits and kind of washed them down to make sure no moon germs got around," says John McLaughlin, one of the *Apollo 11* 'swimmers'.

After getting sprayed and scrubbed with bleach, the astronauts were escorted to a raft, and from there into a net that functioned like a chair (known as a 'Billy Pugh net') and hoisted into a helicopter hovering overhead. They were then sent to quarantine for 21 days. Their families and the eager President Nixon could only talk to them through a barrier. But it wasn't just Nixon and the anxious families, the whole world wanted to touch them. The moon had bought the earth together. The Soviet Union itself accepted its defeat in the Moon race with dignity. The Soviet authorities immediately congratulated the U.S.; the moon landing was shown on Soviet television on the next day and covered in the press. The International Space Station, launched in 1998 would take this blossoming relationship into low earth orbit. It seemed like it worked a lot better out in space.

Life after walking on the moon

For the moon men, life on Earth could only be an anti-climax. Armstrong, Aldrin, and Collins embarked on divergent paths, each influenced by their experience of their journey to the moon. In 1971, Armstrong visited the Soviet Union and as the Soviet Union had previously launched a probe that delivered some moon soil back to Earth, the two countries exchanged the samples. The first human to walk on the lunar surface, then took on a more private role, dedicating himself to teaching as a professor of aerospace engineering at the University of Cincinnati and serving on the boards of several corporations. He remained a thoughtful and somewhat reclusive figure, rarely engaging in public about his lunar experience until his death in 2012. Aldrin, on the other hand, became a vocal advocate for space exploration, pushing for human missions to Mars and beyond.

His post-Apollo life was marked by both achievements and personal challenges, including battles with depression and alcoholism, which he openly addressed, contributing to public awareness about these issues. Aldrin has written several books and continues to be an influential figure in space advocacy, and, as of 2024, is the only one of the three still around. Collins, the command module pilot who orbited the moon while his colleagues walked its surface, pursued a variety of interests post-Apollo. He briefly worked in government before becoming the director of the National Air and Space Museum, significantly contributing to preserving the history of flight and space exploration. Though they followed different paths, each of the *Apollo 11* astronauts contributed to shaping the legacy of space exploration and the ongoing dialogue about humanity's place in the cosmos.

One of Aldrin's more controversial moments came from an interview in 2015, where he discussed a peculiar experience during the *Apollo 11* mission. Aldrin mentioned seeing something out the window on their journey to the Moon that he couldn't identify or explain, leading to speculation and intrigue about UFOs and extraterrestrial life:

> *"There was something out there that was close enough to be observed... sort of L-shaped."*

This comment sparked a flurry of speculation and fueled conspiracy theories about what the *Apollo 11* astronauts might have seen on their historic journey. Aldrin has clarified that his comments were taken out of context and that he does not believe what he saw was an alien spacecraft.

Armstrong—one of the two who walked on the moon—was a strong, silent and brave American hero. Over time, this image became a lot more insular and grouchy: "I don't want to be a living memorial." He didn't seem happy with what happened to the space program after they had walked on the moon. What was it all for? His wife, Janet Armstrong, was quoted as saying "Silence is Neil Armstrong's answer. The word 'no' is an argument. He is a very solitary man."

The least remembered astronaut from the mission, Michael Collins seems to have landed back on Earth a little more easily than his companions. Maybe something did happen out there on the moon to the other two.

> **Collins:** "Neil Armstrong was born in 1930. Buzz Aldrin was born in 1930, and Mike Collins, 1930. We came along at exactly the right time. We survived hazardous careers and were successful in them. But in my own case at least, it was 10 per cent shrewd planning and 90 per cent blind luck. Put 'Lucky' on my tombstone."

In the modern world where attention seeking is a way of life, there's something endearing about the reluctance of these three astronauts to embrace their fame. And perhaps that's why people remain fascinated with them.

CHAPTER 2

But did the moon landing really happen?

Gather round, friends, for a tale that takes us beyond the realms of Earth, past the atmospheric curtain, and onto the surface of the moon—or so they say. Since that fateful day in 1969, when humanity allegedly took its first small steps on lunar soil, a saga has unfolded, not of stars and spaceships, but of scepticism and whispers of deception. Yes, we're talking about the moon landing, or as some would have it, the greatest show not on Earth.

In the aftermath of the Apollo missions, while the world marvelled at what seemed like a leap into the future, a curious counter-narrative began to take shape. It wasn't long before some intrepid souls, armed with nothing but keen eyes and a healthy dose of doubt, started to question the veracity of NASA's claims.

"Look at the flags!" they cried, pointing to footage where the United States flag fluttered as if caught in a lunar breeze. "And the shadows!" they exclaimed, noting how some seemed to defy the logic of a single light source: the sun.

But why, you might wonder, would anyone concoct such an elaborate ruse? Therein lies the crux of our story, dear reader, set against the backdrop of the Cold War, a period rife with espionage, rivalry, and the unyielding quest for technological supremacy. The United States, locked in a cosmic ballet of one-upmanship with the Soviet Union, had much to gain from declaring victory in the space race. What better way to showcase American prowess and ingenuity than by broadcasting a moon landing for the world to see?

The sceptics argue that the technological capabilities of the era simply weren't up to the task. "They could barely get a computer to fit in a room, let alone navigate the vast expanse of space," the doubters jeer, suggesting that the Apollo missions were more Hollywood than Houston. Indeed, the notion that in an age of rotary phones and typewriters, humans could conquer the moon seems, to some, as fanciful as a science fiction novel.

And then there are the photographs—each one scrutinised more closely than a Renaissance masterpiece. Anomalies aplenty have been spotted, fuelling fires of conspiracy that have burned for decades. "Why aren't there any stars?" the sceptics ask, peering into the black void of the lunar sky in the photographs. "How can we trust what we're shown when the very cosmos seems to have been omitted?"

Yet, as captivating as these conspiracy theories may be, spinning a yarn that spans Earth to the moon and back—and beyond—they often buckle under the weight of scientific scrutiny and rational explanation. The fluttering flag? A result of momentum and a lack of air resistance. In fact, the astronauts knew there would be a problem with the floppy flag and inserted a rod to keep it nice and erect. But they didn't quite get it right and it rippled instead. The odd shadows? That's the result of lunar topography and the angle of the sun. The missing stars? A consequence of camera settings not suited to capturing the faint light of distant stars in the bright lunar morning.

But let's indulge, for a moment. The moon landing conspiracy theories, whether grounded in reality or the product of imaginative minds, serve as a testament to our inherent desire to question, to probe the boundaries of our knowledge, and to seek the truth, however outlandish it may seem.

So, as we look up at the night sky, and see the moon that has inspired countless generations, we're reminded of the enduring allure of the unknown. Whether the moon landings were a monumental step for mankind or an elaborate hoax orchestrated for the eyes of the world, they've undeniably etched themselves into the annals of human history, leaving us to wonder, to debate, and to dream of what lies beyond the stars.

Historical context

The inception of the space race can be traced back to the late 1950s when the Soviet Union's successful launch of *Sputnik*, the first artificial satellite, shocked the world and spurred the United States into action.

This event, followed by Yuri Gagarin's historic orbit around the Earth in 1961, demonstrated the Soviet Union's advanced capabilities in space exploration, prompting the United States to respond with an ambitious space program of its own. President John F. Kennedy, recognising the potential implications of falling behind in the space race, set a bold goal for the nation: to land a human on the moon and return them safely to Earth before the end of the 1960s.

The Apollo program became the focal point of the United States' efforts to achieve this goal. It was a testament to the nation's commitment to not only advancing scientific knowledge but also demonstrating its technological superiority and ideological resilience. The program's success required unprecedented levels of innovation and collaboration, leading to significant advancements in rocketry, navigation, and computer technology, among other fields. The moon landing, achieved by *Apollo 11* in July 1969, was the culmination of these efforts.

Despite the persistence of conspiracy theories, extensive evidence supports the authenticity of the moon landing. Independent verification from other countries, analysis of moon rock samples brought back to Earth, and the technological artifacts left on the moon's surface, such as retroreflectors used for laser ranging experiments, provide incontrovertible proof of the Apollo mission's success. In addition, advancements in technology and the dissemination of information have allowed for a more thorough debunking of hoax claims, reinforcing the moon landing's status as a genuine historical event.

Scepticism and conspiracy theories

While the majority of the world celebrated this unprecedented feat of the moon landing in 1969, a segment of the population began to harbour and propagate doubts, leading to the genesis of various conspiracy theories. These theories proposed that the moon landing was not an authentic milestone in human exploration, but a sophisticated hoax crafted by the U.S. government and NASA. The motivations behind these assertions were multifaceted, deeply entwined with the sociopolitical fabric of the time, and the technological landscape, which, to some, seemed insufficiently advanced for such an endeavour.

The period following the moon landing was ripe for the development of scepticism and conspiracy theories. The Cold

War, a backdrop of intense rivalry between the United States and the Soviet Union, was marked by espionage, propaganda, and a fierce competition for technological supremacy. In this climate of distrust and competition, the moon landing was viewed by some as too opportune, a perfect embodiment of the desire of the United States to assert its dominance on a global stage. These doubts were compounded by the fact that the space race was as much a political endeavour as it was a scientific one and, because of this, the moon landing's role as a political statement, alongside its scientific achievements, provided fertile ground for those inclined to question its authenticity.

The lack of a direct external verification of the landings at the time, given that the United States was the only country capable of sending missions to the moon, added layers of doubt. This scepticism was not dampened by the broadcasts of the landing, which were viewed by millions worldwide: instead, the very scale of the broadcast and the flawless execution of the mission as presented to the public only fuelled further speculation and disbelief among sceptics.

As the years progressed, these conspiracy theories found a foothold in popular culture, bolstered by books, documentaries, and the ubiquitous internet that allowed for the rapid dissemination of ideas. The theories evolved and became more intricate, incorporating aspects of emerging technology and media to argue that NASA and the government had the means to fabricate the entire mission. Bill Kaysing in 2001 (one of the original moon conspiracy theorists), author of *We Never Went to the Moon: America's Thirty Billion Dollar Swindle* (1976):

> *"The astronauts were launched with the Saturn V. Then, in order to account for their disappearance, they simply orbited the Earth for eight days and in the interim they showed these*

fake pictures of the astronauts on the moon. But on the eighth day, the command console separated from the vehicle and descended to Earth as, of course, was shown in the films."

It's actually quite hard to find people who admit to these theories. Podcaster Joe Rogan was originally one of them but seems to have changed his mind. As far as true believers go, in 2019, 5 per cent of the U.S. population believes the moon landing was faked. That's quite a few people—more than 16 million! In that same survey, 61 per cent strongly disbelieved it was faked, leaving quite a large portion that seem to be a bit undecided. Joe Rogan's huge audience are obviously fascinated by this sort of thing. Aren't we all?

However, as much as these theories are entertaining, extensive evidence supports the authenticity of the moon landing. Advancements in technology over the decades have allowed independent observers to verify aspects of the missions, such as imaging the landing sites and the equipment left behind.

The development of moon landing conspiracy theories reflects a broader scepticism towards government and authoritative bodies, a scepticism that has been exacerbated by genuine instances of misinformation and political manipulation throughout history. While the evidence overwhelmingly supports the reality of the moon landing, the persistence of these theories speaks to the complexities of public trust, the interpretation of technological achievements, and the desire for a more transparent understanding of historic events.

Anomalies in photographs and videos

The visual documentation of the 1969 moon landing has been scrutinised almost as much as it has been celebrated. Conspiracy-minded individuals have pointed to what they perceive as inconsistencies and anomalies in the photographs and videos provided by NASA as evidence that the moon landing was, in fact, a staged hoax. The development of these conspiracy theories, particularly focusing on the anomalies in visual documentation, has fuelled a debate that persists decades after the event.

Among the most cited anomalies is the "behaviour" of the American flag, which appears to wave in the vacuum of space. Sceptics argue that this motion suggests the presence of wind, which would be impossible in the moon's atmosphere-less environment. This singular observation became a cornerstone for conspiracy theorists, who suggested that the flag's movement was proof of on-Earth filming, where air movement caused the flag to flutter.

Additionally, the lighting and shadows in the moon landing photos and videos have been subjects of intense scrutiny. Conspiracy theorists contend that the lighting is too even for the images to have been taken in the stark, unfiltered sunlight of space. They suggest that multiple light sources were used, akin to a film set, rather than the single light source of the sun. They argue that shadows cast by objects in some photographs appear to intersect rather than run parallel, as would be expected under a single light source. These observations are presented as evidence of artificial lighting, supporting claims that the landing was staged and filmed on a set.

The proliferation of these conspiracy theories was facilitated by a combination of factors. First, there was a general lack of public understanding of the scientific and technical complexities involved in space exploration. The nuances of how objects behave in space, the effects of sunlight without Earth's atmospheric diffusion, and the properties of materials used in space missions were not common knowledge and this gap in understanding created a fertile ground for scepticism and speculation.

The response to these conspiracy theories has been multifaceted. Experts in photography, physics, and space science have provided explanations for the anomalies noted by sceptics. The apparent waving of the flag, for instance, is attributed to the flagpole's design, which included a horizontal rod meant to keep the flag extended. Any movement seen in photographs was due to the astronauts' handling of the pole and rod, not wind. As for the lighting and shadow anomalies, these have been debunked through demonstrations that replicate the lunar surface's conditions, showing that complex shadow interactions can indeed occur under a single light source due to the moon's terrain and the angles of light and reflection.

Despite these explanations, the conspiracy theories surrounding the moon landing's visual documentation have persisted, evolving with time and technology, especially through the internet, which has played a significant role in both spreading and debunking these theories, serving as a platform for discussion, analysis, and the exchange of information.

Technological capabilities

Central to these numerous conspiracy theories is the doubt cast upon the technological capabilities of the 1960s, with sceptics questioning whether the era's technology was sufficiently advanced to accomplish such a feat. This scepticism not only challenges the authenticity of the moon landing but also invites a deeper exploration into the nature of technological innovation and its perception by the public.

During the 1960s, the space race pushed both superpowers at the time—the Soviet Union and the United States—to accelerate their space exploration efforts. Despite this, the notion that the technology of the time could facilitate a manned lunar mission, complete with a safe return to Earth, seemed audacious to many. This incredulity was partly rooted in a limited public understanding of the scientific principles and technological innovations at play, as well as the extraordinary pace at which advancements were being made.

Critics of the moon landing often point to various aspects of the mission that they believe could not have been feasible with 1960s technology. These include the survivability of astronauts through the Van Allen radiation belts, the functionality of the lunar module in the moon's gravity and vacuum, and the reliability of the computers and navigation systems necessary for such a mission. The doubt was compounded by the fact that, just a decade prior, space travel was more a matter of science fiction than scientific planning.

But there were plenty of computers involved in the Apollo missions and they were a crucial part of its success.

> *17 July 1:15:56 hrs somewhere between the earth and the moon...*
> **Mission Control Center, Houston:** *Yes, Mike. We show you in VERB 59 right now. Over.*
> **Command Module Pilot:** *That's right. I...I haven't entered ... I gave the ... I gave it back to the computer for a second. I put the mode switch from MANUAL back to CMC while I fooled with the DSKY, and the computer drove the star off out of sight. So the delay here has been in going back to MANUAL and finding the star again, which I've finally done. And... just a second here, I'll go to ENTER and get a 51 and mark on it. As I say, for some reason the computer drove the star off out of sight.*
> **CC:** *Okay. Roger. Out.*
> **CC:** *Apollo 11, this is Houston. Over.*

A faked transcript perhaps, but what is often overlooked by proponents of the hoax theory is the unprecedented investment in research and development that the Apollo program represented. The program benefited from the mobilisation of vast resources and the collective efforts of some of the brightest minds in engineering, physics, and computer science. This collaboration led to significant technological innovations, including the development of the *Saturn V* rocket, the most powerful rocket ever built, capable of propelling the Apollo spacecraft beyond Earth's orbit; the lunar module, designed specifically for moon landings; and the Apollo guidance computer, one of the first to use integrated circuits, providing the computational power necessary for navigation and control.

In addition, the technologies and materials developed or refined during the Apollo program, such as heat shields capable of withstanding the extreme temperatures of re-entry and lightweight, durable materials used in the astronauts' space suits, were cutting-edge for their time. These advancements were not just sufficient for the task at hand but represented significant leaps forward in their respective fields.

The conspiracy theories that question the technological feasibility of the moon landing also tend to underestimate the scale and scope of the Apollo program. The effort was not a sudden leap into the unknown but rather the culmination of years of incremental progress, testing, and learning from earlier missions. Each mission in the Apollo series built upon the successes and failures of its predecessors, refining technologies and techniques in a deliberate progression toward the goal of a manned lunar landing.

The persistence of conspiracy theories reflects not so much the inadequacies of 1960s technology but rather the challenges of understanding and accepting the rapid pace of technological advancement, especially when such achievements touch upon the collective aspirations and fears of an era. Yes, it was a technically difficult operation. But not impossible. It still stands as one of humanity's greatest achievements.

Political motivations

President Kennedy's bold declaration that the United States would "send a man to the moon" and bring him safely back to Earth by the end of the decade was as much a statement of technological ambition as it was a political declaration aimed at restoring national pride and demonstrating American resilience and capability.

As a result, the moon landing was imbued with profound political significance. It was not merely an exploration or scientific endeavour but a symbol of victory in the Cold War's ideological and technological battleground. This dual nature of the moon landing has been a central theme in the development of conspiracy theories and sceptics have long argued that the immense pressure to surpass the Soviet Union in space exploration could have motivated the United States government and NASA to stage the moon landing as a propaganda tool. The argument follows that the potential embarrassment of failing to achieve Kennedy's goal, coupled with the strategic necessity of demonstrating global leadership, made the moon landing too important to leave to chance. So they had to make the TV show!

These conspiracy theories have been further fuelled by the secretive nature of the Cold War, where misinformation and psychological operations were common tools used by both sides to sway public opinion and international perception.

1. Conspiracies can now be searched anywhere and everywhere. 2. Bill Kaysing with one of the moons he claims were never set foot on. 3. The workplace of a typical moon conspiracy theorist. 4. Footprints from the moon. Allegedly? 5. The White House and the full moon. 6. 'Buzz' Aldrin on the moon... but really?

In an era defined by distrust, the leap from scepticism about the authenticity of the moon landing to belief in a comprehensive hoax is shorter than it might otherwise be. The staging of such an event, according to conspiracy theorists, would serve not only to claim technological superiority over the Soviet Union but also to bolster domestic and international perception of American power and ingenuity.

The Cold War context has lent a certain plausibility to these conspiracy theories for those predisposed to doubt official narratives. The existence of actual propaganda and covert operations has provided a historical basis that conspiracy theorists use to bolster their claims, suggesting that staging a moon landing would be within the realm of possibility for a government engaged in a global ideological struggle.

This line of reasoning overlooks the vast body of scientific evidence and international verification that supports the authenticity of the moon landing, instead focusing on the political motivations that could underlie such an endeavour.

Independent verification

Amidst claims that the landings were staged by the United States government and NASA, a critical body of evidence often overlooked in these discussions comes from independent scientific verifications. These verifications have played a central role in substantiating the authenticity of the moon landings, offering a counterpoint to the scepticism and serving as a linchpin for rational discourse on the subject.

One of the most compelling pieces of evidence supporting the reality of the moon landings is the presence of retroreflectors on the lunar surface. These devices, which were placed there by Apollo astronauts, are designed to reflect laser beams sent from Earth. Scientists and astronomers, not affiliated with

NASA, have used these retroreflectors for experiments that measure the distance between the Earth and the moon with remarkable precision. This ongoing experiment, which can be performed by independent observatories around the world, unequivocally confirms that human-made objects have been placed on the moon, directly contradicting claims that the moon landings were fabricated.

The international community of astronomers, leveraging telescopes and laser-ranging equipment, has also independently observed and verified the results obtained through interactions with the lunar retroreflectors. The physics underlying these experiments is well-documented and universally accepted, further cementing the veracity of the moon landings.

These observations offer a scientific method for verification that transcends political and nationalistic biases, grounding the discussion in empirical evidence and rational analysis.

Scientific analysis of moon rocks

The Apollo moon missions culminated in six manned landings between 1969 and 1972, and not only achieved the unprecedented feat of placing humans on the lunar surface but also returned with a wide selection of lunar samples. These moon rocks, meticulously collected and brought back to Earth, have become a focal point of scientific study and a significant counterargument against the conspiracy theories that claim the moon landings were hoaxed. The analysis of these lunar samples by scientists around the world has yielded insights into the moon's composition, formation, and history, offering conclusive evidence of their extra-terrestrial origin and, by extension, the authenticity of the Apollo missions.

Upon their return to Earth, the lunar samples were distributed to scientific institutions globally to undergo rigorous examination. This widespread dissemination ensured that the analysis was not confined to American scientists or NASA-affiliated laboratories but involved a broad international scientific community. The findings from these analyses have been remarkably consistent, revealing that moon rocks possess unique chemical compositions and physical properties that distinctly set them apart from any terrestrial rocks. For instance, the ratio of isotopes, such as oxygen isotopes, in lunar samples is markedly different from those found in Earth rocks, indicating their non-earthly origin.

Moon rocks lack certain minerals that are ubiquitous on Earth's surface and contain others that, while present on Earth, form under conditions not found on our planet. This disparity lends weight to the argument that these samples could not have been clandestinely collected or fabricated on Earth, as some conspiracy theorists suggest. The rocks also exhibit features indicative of their formation in the vacuum of space, such as the presence of "zap pits" caused by micrometeorite impacts, something that cannot be replicated in Earth's atmosphere.

Another compelling piece of evidence comes from the age of the lunar samples. Radiometric dating techniques have shown that some of the moon rocks are up to 4.5 billion years old, significantly older than the majority of surface rocks on Earth, which have been continually recycled through geological processes. This age difference provides a timeline for the moon's history that aligns with current scientific understanding of the solar system's formation and evolution.

The argument that the moon rocks could have been meteorites from the moon that landed on Earth and were subsequently presented as collected from the moon during the Apollo missions has also been debunked. The specific conditions under which the samples were collected, documented, and sealed on the moon negate the possibility of terrestrial contamination. In addition, the sheer volume of samples returned—382 kilograms (842 pounds) over six missions—far exceeds the quantity and variety of lunar meteorites found on Earth. The moon rocks returned by the *Apollo* astronauts serve as a tangible link to the moon, offering incontrovertible evidence of humanity's presence there.

Testimonies from astronauts and space scientists

The testimonies of astronauts who walked on the moon and the scientists involved in the Apollo missions stand as powerful affirmations of the moon landings' authenticity. These firsthand accounts and expert insights provide not only a rebuttal to the conspiracy theories but also a humanising perspective on one of humanity's greatest achievements.

The astronauts who participated in the Apollo missions, including the twelve who walked on the moon, have consistently and unequivocally affirmed the reality of their experiences. Their detailed accounts of the lunar surface, the challenges they faced, and the profound impact of their journeys offer compelling evidence that these missions were genuine. These testimonies are not merely anecdotal; they are backed by extensive training, rigorous scientific methodology, and the physical evidence collected during the missions, including photographs, videos, and lunar samples.

Beyond their personal experiences, the astronauts have also spoken about the broader implications of the Apollo missions. They reflect on the spirit of exploration, the technological and scientific advancements achieved, and the collective effort of thousands of individuals who contributed to the success of the moon landings. Their stories underscore the complexity and authenticity of the missions, challenging the simplistic and unfounded claims presented by conspiracy theories.

Jim Lovell, an astronaut who flew to the moon twice during the Apollo missions, came out against the conspiracy theorist Bill Kaysing, saying "the guy is wacky. His position makes me feel angry. We spent a lot of time getting ready to go to the moon. We spent a lot of money, we took great risks, and it's something everybody in this country should be proud of.' Kaysing responded in the proud American way by suing him for slander. The case was dismissed in 1997.

In addition, the astronauts, scientists and engineers involved in the Apollo missions have provided critical insights into the technological and scientific underpinnings of the moon landings. These experts have detailed the development of the *Saturn V* rocket, the lunar module, the command module, and the various scientific instruments used during the missions. They have also explained the rigorous processes of planning, simulation, and problem-solving that were integral to the success of each mission.

Technological evidence

One of the most compelling pieces of technological evidence comes from the advancements in imaging technology. High-resolution photographs taken by lunar orbiters in the years following the Apollo missions have captured the landing sites with stunning clarity. These images show not only the descent stages of the lunar modules that remain on the moon's surface but also tracks left by the astronauts and the rovers in the lunar soil. The precision with which these artifacts can be identified and matched with the historical mission logs serves as powerful evidence against the hoax theories. The capability to capture such detailed images from lunar orbit is a testament to the strides made in imaging technology, allowing for scrutiny of the moon's surface that was not possible in the immediate aftermath of the Apollo missions.

Further bolstering the case for the moon landings' authenticity are advancements in satellite technology and remote sensing. Satellites equipped with cameras and sensors capable of capturing images in various spectra have provided additional layers of evidence. For instance, the lunar reconnaissance orbiter, launched by NASA, has conducted extensive mapping of the moon's surface, offering high-resolution images that corroborate the locations of the Apollo missions. The ability of these satellites to precisely locate and image objects on the moon's surface from orbit provides a direct counter to claims that the landings were fabricated.

The sheer magnitude of technological evidence that has emerged in the years following the Apollo moon landings serves as a robust foundation for affirming their authenticity. The advancements in imaging technology, satellite remote sensing, laser ranging, and digital forensics have not only debunked the hoax theories but also enriched our knowledge and appreciation of the moon.

Global consensus

The Apollo moon landings between 1969 and 1972 were huge achievements for humankind and these missions have been widely recognised and celebrated across the globe. The overwhelming consensus among the scientific community and the general global population is that the moon landing is a factual historical event, with these conspiracy theories largely discredited through a combination of scientific evidence, technological verification, and logical scrutiny. The credibility of the moon landings is also reinforced by the transparency and openness with which NASA has conducted its space exploration endeavours. The Apollo missions were broadcast live, allowing millions around the globe to witness these historic events in real time. The astronauts who participated in these missions have shared their experiences extensively, participating in public speaking events, interviews, and educational programs. Their firsthand accounts, along with the vast archive of visual, audio, and written documentation of the missions, provide a coherent and consistent narrative that corroborates the historical fact of the moon landings.

The acceptance of the moon landing as a factual historical event by the scientific community and the global population at large stands as a testament to the power of scientific inquiry, technological advancement, and human ambition and it underscores the importance of critical thinking and evidence-based reasoning in distinguishing fact from fiction. But despite the volumes of evidence and overwhelming information that is available to the public, there is still a sizeable portion of the population—nowhere near a majority but still a large number—that question the authenticity of this material.

Why do the conspiracies continue?

The persistence of conspiracy theories regarding moon travel, particularly the claims that the Apollo moon landings were hoaxed, presents an intriguing paradox in the face of overwhelming evidence to the contrary. If people can still believe this—what else are they capable of believing? This phenomenon raises critical questions about the nature of belief, scepticism, and the factors that fuel the acceptance of conspiracy theories despite all evidence. The reasons behind the endurance of such theories are multifaceted, involving psychological, sociological, and media-related dynamics. It is neither accurate nor fair to categorically label individuals who accept these conspiracies as having mental health issues or being merely gullible; instead, a careful exploration reveals a complex range of factors that contribute to the belief in such theories.

One of the primary psychological factors at play is cognitive dissonance, the discomfort experienced when holding two conflicting beliefs or when new information contradicts existing beliefs. For some, the monumental achievement of moon travel conflicts with their perception of technological capabilities, leading them to reject this reality in favour of a conspiracy that aligns with their preconceived notions of possibility. This mechanism is compounded by confirmation bias, where individuals favour information that confirms their pre-existing beliefs and dismiss information that contradicts them. The internet and social media have amplified this bias, enabling individuals to find communities and sources that reinforce their views, regardless of the factual accuracy of these beliefs.

Another psychological aspect is the propensity for pattern recognition and agency detection, inherent traits that have evolutionary benefits but can lead to the erroneous identification of patterns or intent where none exist. In the context of moon landing conspiracies, anomalies in photographs or narratives are interpreted as deliberate manipulations, leading to the belief in a hoax despite logical explanations for these observations.

Sociologically, conspiracy theories often emerge in times of societal stress, uncertainty, or distrust in institutions. The moon landings occurred during the Cold War, a period marked by high stakes geopolitical tension and widespread distrust of government narratives. In contemporary times, this distrust has persisted and even expanded, fuelled by real instances of misinformation and deception. For some, questioning the moon landings is an expression of broader scepticism towards authority and a manifestation of the desire to challenge perceived narratives of power.

It is also crucial to consider the role of community and identity in the belief in conspiracy theories. For some individuals, these beliefs are part of a larger worldview that offers a sense of belonging and identity within a community that shares these views. This social aspect can reinforce belief in the conspiracy, as acceptance and agreement within the community are prioritised over external evidence or mainstream narratives.

Addressing the question of whether individuals who accept moon landing conspiracies are correct requires a return to the evidence. The scientific consensus, based on empirical data, technological artifacts, and independent verification, overwhelmingly supports the reality of the moon landings. The burden of proof lies with those claiming a hoax, and to date, no evidence has been presented that withstands scientific scrutiny and rational examination.

A strawpoll inquiring about the moon landing usually ends up with an affirmative answer. Anyone who is asked: "Do you really think that governments are that organised to create a fictitious moon landing? The entire world was watching on television"—will usually respond with "Yes, it happened," and this response is across all age groups, even for the people who were born way after 1969. Yet some scepticism still resides. One reason the scepticism remains is the fact that humans have not walked on the moon since 1972. The real question is: what happened to the moon program? Did we discover that all there was up there was just a pile of worthless moon dust?

★

CHAPTER 3

They're back! A return to the moon

A lot of things made by humans will be aimed at the moon in 2024. This year will see an international flotilla of missions—each with its own unique objectives—embarking on a voyage to our nearest celestial neighbour. With agendas ranging from scientific inquiry to techno showboating, these missions aim to prepare the lunar dust for future generations of explorers. Some of this may or may not happen, as the history of space has proven. Indeed, early in the year there has already been one failure—a leaky fuel tank has prevented the privately funded *Peregrine* moon lander from... landing.

It was a somewhat controversial mission as the lander was carrying a payload that included capsules of incinerated human remains. These included science fiction writer Arthur C. Clarke; *Star Trek* creator Gene Roddenberry; Roddenberry's wife, Majel Barrett; and Nichelle Nichols, James Doohan and DeForest Kelley, Montgomery Scott and Dr. Leonard McCoy, all actors from this classic sci-fi show. Stored alongside these remains were samples of DNA of the U.S. presidents George Washington, Dwight Eisenhower, John F. Kennedy, and

Ronald Reagan, destined to become one with the moon dust. But it was not to be.

A spokesman for the Navajo Nation, Justin C. Ahasteen, the organisation's Washington executive director, told the media.

"While we are relieved that the Peregrine mission failure means that the lunar surface will not become a resting place for human remains, we recognise the disappointment and setback this represents for all those involved in the mission… Our concerns are specific to the inclusion of human remains on missions intended for the moon, which we consider sacred."

The coalition for lunar exploration

In what can be described as a renaissance of lunar exploration, countries around the globe are pooling their resources and expertise. The United States' NASA, Europe's ESA, Japan's JAXA, and India's ISRO, along with a phalanx of commercial partners, have outlined ambitious plans for a series of interlinked missions that could transform our understanding of the moon. Early in 2024 Japan has become the fifth nation to make a soft lunar landing with its robotic SLIM spacecraft. Here's a quick look at some other moon things. We wish them all well.

The cornerstone of the 2024 missions is the Lunar Resource Mapping Mission. A fleet of satellites equipped with advanced spectrometers will scan the moon's surface to chart its mineral wealth, seeking out iron, titanium, and potential water ice reserves. Their goal is to map the distribution of valuable minerals like iron and titanium, and to identify deposits of water ice. These resources are critical for supporting long-term human presence on the moon by providing raw materials for construction, life support, and potentially fuel.

The data collected will not only advance scientific knowledge but also inform the development of technologies for resource extraction and utilisation on the moon. This mission is about turning scientific data into practical information that can make lunar settlement possible.

These satellites equipped with spectrometers will not only map existing resources but also potentially unveil new ones. The spectrometers, sensitive to a wide range of wavelengths, will detect the unique spectral fingerprints of various minerals and ices. By creating a detailed resource atlas of the moon, this mission will guide future decisions on where to land, where to build, and where to mine. This data will also help to optimise the design of life support systems, habitats, and manufacturing processes that utilise lunar materials, making human presence on the moon more sustainable and economically feasible.

The moon doesn't look very wet. But Dr. Ana Rivera, Lead Scientist for the initiative says: "Water is a game-changer—it can support life and be converted into rocket fuel. Our spectrometers are fine-tuned to detect its presence. This mission's findings will lay the groundwork for the next giant leap in space travel and permanent lunar settlements."

Dr. Rivera's team is meticulous in their approach, knowing that the success of future lunar endeavours hinges on the data they gather. The spectrometers onboard these satellites are the most advanced yet, capable of detecting subtle variations in the lunar soil that could indicate the presence of vital resources. "It's not just about going back to the moon; it's about staying there. Our work will help ensure that when humans return to the moon, we'll be there to stay."

Her positive spin is reminiscent of the 1970's moon fever that swept the globe. But there's more.

The pioneers: roving the lunar wastelands

A new fleet of lunar rovers from various space agencies and companies will soon navigate the moon's surface. These rovers are equipped with scientific tools and drills to search for water and study the moon's environment. Their findings will help determine if the moon can support human life in the future.

The upcoming lunar missions, featuring NASA's *VIPER* rover and a range of international and commercial partners, are a landmark in space exploration, combining the strengths of government agencies and private organisations. This collaboration is the result of the need to share resources and manage the high costs associated with space missions. Space exploration is notoriously expensive, and by working together, these entities can pool their financial and technological resources. This not only makes the missions more feasible but also allows for a more efficient use of funds and equipment.

Bringing together the expertise of government agencies and private companies enhances the scientific research conducted on the lunar surface. Government agencies like NASA bring years of experience in space missions and a wealth of scientific knowledge, while private companies contribute innovative technologies and new approaches. And, they can fire people more easily. This synergy expands the scope and depth of the lunar exploration, allowing for more comprehensive studies and experiments.

There are also significant economic and strategic reasons behind these collaborations. Governments view lunar missions as opportunities to access new resources and gain strategic advantages. The moon's potential for rare minerals and water ice can be crucial for both space-based and Earth-based applications. For private companies, involvement in these missions represents a chance to be at the forefront of space industrialisation, opening up new markets and driving innovation in space technology.

The involvement of private partners in these missions is also a step towards the commercialisation of space. This initiative is expected to spur economic growth, creating new jobs and markets in the space industry. It also aligns with the long-term goals of establishing human habitation on the moon. These collaborative missions serve as vital research and testing grounds for technologies and systems required for sustained human presence in space.

International cooperation in these missions embodies a collective effort that goes beyond individual national interests—it brings together diverse perspectives and expertise from around the world, fostering a spirit of global partnership in pursuit of shared scientific and exploratory goals. This not only advances our understanding of the moon but also promotes peaceful international relations and cooperation in space exploration.

Could this be an example of a new peaceful coexistence between these working collectives, while on Earth, the battle rages between pro-government and anti-government believers? This partnership is crucial for exploring new frontiers in space and achieving ambitious goals that would be challenging to accomplish by any single organisation or country.

The Lunar Communication Link

As these rovers and other instruments relay a bounty of scientific data, they'll rely on the Lunar Communication Link Initiative. This network of lunar satellites aims to provide a robust communications relay system, facilitating high-speed data exchange between Earth and the moon.

A robust communication system also lays the groundwork for future lunar exploration and long-term missions. For example, as humans prepare for a return to the moon and potentially establish a permanent presence there, they will need a reliable way to communicate with Earth. This network will serve as the backbone for future navigation and communication systems, which astronauts will depend on for both safety and scientific tasks.

It will likely involve a series of orbiters equipped with the latest technology to handle the unique challenges of space communication. The satellites will need to contend with the moon's harsh environment, the varying distance between the Earth and the moon, and the need to transmit large amounts of data at high speeds.

Then there's the telescopes. On the moon's far side, shielded from Earth's chatter, a network of telescopes will open their lenses to the cosmos. These silent sentinels are expected to peer deeper into space than ever before, thanks to the moon's lack of atmosphere and the absence of radio interference from Earth.

These aren't your typical garden-variety telescopes; they are the interstellar equivalent of deep-sea explorers, poised to dive into the cosmic depths. With the moon's stark landscape as their base, they'll peek into the universe's elusive corners, unobstructed by the atmospheric fuzz and radio cacophony that plagues their Earth-bound cousins. They might even give us a clue to what the invisible dark matter is that seems to hold the universe together.

The moon base

There's also an attempt to erect a kind of display village in the form of a proto moon base. The Lunar Habitat Demonstration mission is a practical step toward sustainable human presence on the moon. It's all about constructing a prototype base that could show how future astronauts might live on the lunar surface. The key here is the use of lunar materials in building the habitat. This approach aims to significantly reduce the need to transport materials from Earth, using the moon's resources instead to build and maintain the structure. It's an exercise in efficiency and self-sufficiency, focusing on the direct application of moon resources to support long-term exploration.

The implications of this mission are straightforward and significant. A successful trial of the habitat could lead to the development of long-term living quarters on the moon, setting the stage for more ambitious projects such as larger bases or even as a launchpad for missions further into the solar system. The goal is to create a sustainable and potentially scalable living environment, using innovative construction techniques such as robotics and 3D printing to leverage the moon's own soil. Ultimately, this mission is a crucial experiment in overcoming the challenges of extra-terrestrial construction and living. Who cares if everything is coloured moon beige?

Getting the kids early

The Cultural and Educational Outreach Mission sounds a bit like something that might have been achieved over 50 years ago, when books like *You Will Go To The Moon* were on many children's bookshelves and this writer was clustered in a classroom watching a black and white television in suburban Australia, which was broadcasting the first landing in 1969.

This century's version comes with a "particular focus on the younger minds back on Earth. It's a mission that bridges the gap between the classroom and the cosmos, transporting the creative and scientific outputs of students to the moon's surface". The idea is a bit more interactive than watching the television transmission of the original lunar landing. By sending experiments and art projects to the moon, the mission aims to ignite the imaginations of young people, showing them that their ideas and creations can quite literally go beyond worldly boundaries. It's about giving students a tangible connection to space exploration, making it more than just a distant concept.

It is a strategic move to foster long-term interest in the fields of S.T.E.M.—science, technology, engineering and mathematics. When students see their projects sent to the moon, the message is clear: space is within reach, and so are careers in science, technology, engineering, and mathematics. The projects that make the lunar journey are not just homework; they are proof of concept for the students' ingenuity. By involving students directly in the lunar missions, the mission creates "a powerful narrative that the next giant leap for mankind could come from the minds of those still in school today".

As art and experiments crafted by students find their home on the lunar surface, the mission weaves the dreams of Earth's children into the fabric of humanity's lunar narrative, fostering a connection with the moon that will resonate for years to come.

Finally: here come the humans... again

Towards the end of 2024, NASA's *Artemis II* mission will see the return of astronauts to the moon—or, at least, close to the moon—after nearly half a century since Apollo's last voyage. Scheduled for no earlier than November 2024, this eight-day mission is charged with significant expectations and excitement. The mission's blueprint involves sending four astronauts on a path that will loop around the moon, encapsulating both a nod to the past and a firm step towards the future of space exploration. The astronauts will be aboard

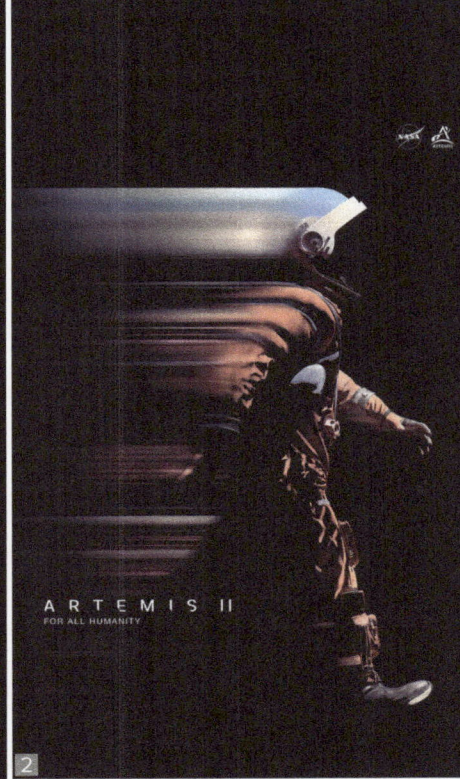

1. Earth and moon. 2. NASA's promotional material for the *Artemis II* project. 3. The Gateway modular space station in lunar orbit. 4. The original cover from *You Will Go To The Moon*, released in 1959. As it turned out, it's still way ahead of its time.

the *Orion* spacecraft, a capsule designed for deep space missions, which will be perched atop the formidable space launch system rocket—the most powerful rocket ever built, representing the pinnacle of current space travel technology.

The *Artemis II* mission serves as a critical testbed for NASA's ambitious lunar exploration plans. It's part of the broader Artemis program, which aims not only to return humans to the moon but also to land the first woman and the next man on the lunar surface by 2025 with the subsequent *Artemis III* mission. *Artemis II* will be pivotal in testing the endurance and capabilities of the *Orion* spacecraft and SLS rocket on a crewed mission, providing data on how the spacecraft performs in the deep space environment. This includes assessing life support and control systems that are crucial for ensuring the safety and success of longer-duration human missions.

Artemis II is expected to lay the groundwork for international cooperation in space. While the mission is a NASA project, it's supported by an international consortium, with contributions from the European Space Agency (ESA) such as the European Service Module, which will provide *Orion*'s power, propulsion, and life support. This collaboration signifies the global nature of future space exploration efforts. Through *Artemis II*, NASA seeks to demonstrate the possibilities of international partnerships in achieving common goals in space, setting the stage for sustainable exploration with astronauts from around the world.

Looking forward

As they say in the corporate jargon of this century… 'looking forward'. Will this new wave of moon activity bring us any closer to living and loving on the moon? In the 1970 book *The First On The Moon*, the original American astronauts Neil Armstrong,

Michael Collins and Buzz Aldrin tell their incredible story with the help of a ghost writer and copious transcripts from the recordings NASA made of their transmissions to earth. The end of the book contains an epilogue by the English science fiction writer and futurist Arthur C. Clarke, where he makes some predictions.

"Anything written about the moon at the beginning of the 1970s will probably look silly in the 1980s and hilarious in the 1990s—particularly to the increasingly numerous inhabitants of our first extra-terrestrial colony." Although some of his predictions have become reality—including satellites, communications and something he predicted that sounds a lot like the internet, he goes on to assure us the first baby will be born on the moon "well before the end of this century", although "it would be interesting to know the nationality of its parents". Will any of these things happen by the end of this new century? We will not make any more predictions but 2024 could prove to be a crucial year. Or maybe 2025 when humans are meant to return to the moon in person on *Artemis III*. But it's safe to say, anything could happen. The Russians might like to reassert their space skills, the Chinese are busy doing something and one of those tech billionaires might beat them all.

CHAPTER 4

The astronomical and geological history of the moon

Imagine gazing up at the night sky, where the moon hangs like a luminous orb, its surface a massive horizon of light and shadow. Earth's closest companion in the vast expanse of space, holds within its craters, mountains, and plains, a story not just of itself but of our solar system—a narrative spanning over 4.5 billion years. Let's go on a journey from the moon's dramatic birth and geological evolution through to its role in the solar system's celestial mechanics. It's a story that all those moon rocks gathered on the Apollo missions have helped to tell.

Our story begins with a cataclysmic event, a colossal collision between the early Earth and a Mars-sized wanderer, which led to the moon's formation. Imagine the chaos, the energy, the sheer scale of an impact that would create a new celestial body from the debris. This violent event set the stage for the creation of something extraordinary. In the aftermath, the moon was a molten sphere, its surface roiling with an ocean of magma, a planetesimal undergoing the throes of differentiation to forge

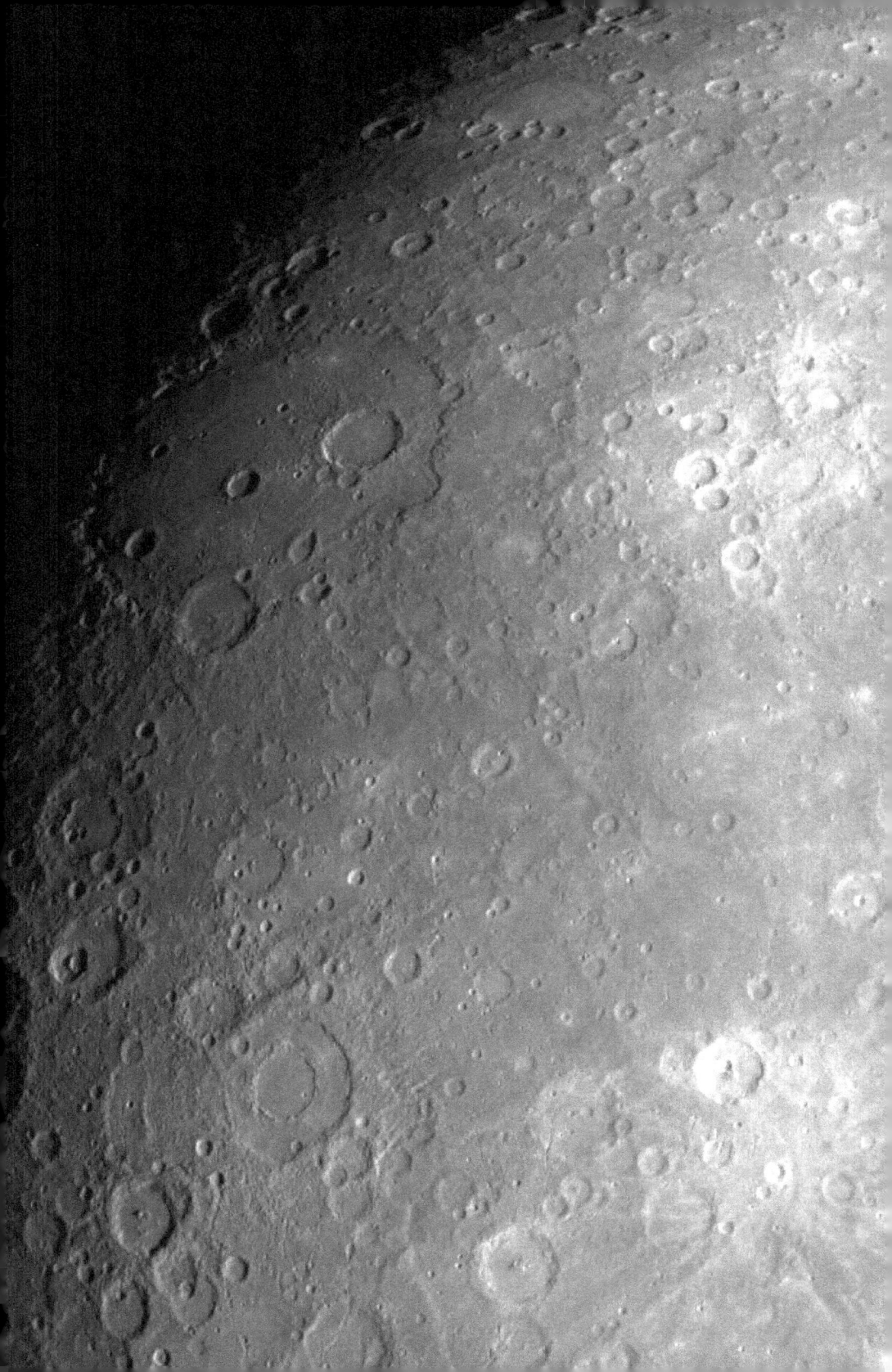

its crust and mantle. As time passed (as in, quite a few eons), the moon's visage was sculpted by a relentless barrage from the cosmos—meteorites that pocked marked its face, creating the cratered appearance we see through telescopes. Each crater, a mark of the solar system's turbulent history, tells a story of creation and destruction. Amidst this landscape of craters rise the maria and highlands, stark in their contrast and beauty, revealing tales of volcanic eruptions that painted vast plains with dark basalt and sculpted towering mountains.

Beneath the footprints of the Apollo astronauts lies the regolith, a fine, powdery testament to billions of years of meteoritic impacts. This dusty layer holds keys to understanding not just the moon, but the processes that govern our solar system. The rocks collected by these astronauts on their lunar travels—a treasure trove of geological clues—have unveiled the moon's composition, shedding light on its formation and the myriad forces that shaped its surface.

But the story of the moon is not just one of passive change—it has witnessed tectonic shifts, evidence of a dynamic past, etched into its surface in the form of lobate scarps and wrinkle ridges. These features speak to a time when the moon, though not tectonically active like Earth, was nonetheless a world in flux.

Through the lens of the Apollo missions, we have glimpsed the moon's role in the grander narrative of solar system formation. The lunar rocks, silent witnesses to the conditions and processes of an earlier epoch, offer insights into the birth of planets and the early solar system. They tell us that the moon is more than Earth's satellite; it is a key to understanding the astronomical evolution of celestial bodies.

Let's explore the moon not just as an object in the night sky, but as a world with its own complex history. It is a story of origins and transformations, of collisions and eruptions, a narrative that enriches our knowledge of planetary science and the intricate dance of the cosmos. Welcome to the journey through the astronomical and geological history of the moon. It's full of dust, explosions, volcanoes and odd terms like "lobate scarps". Let's go!

Formation

The prevailing theory about the Moon's origin, which suggests that it was born from the cataclysmic collision of a Mars-sized body with the early Earth, offers a compelling narrative that not only explains the formation of the Moon but also sheds light on the early dynamics of our solar system. This event, occurring approximately 4.5 billion years ago, marked a definitive moment in the nascent stages of the Earth–moon system and has left an indelible mark on both bodies.

In the aftermath of this colossal impact, a significant amount of debris was ejected into orbit around the Earth. The energy released by the collision was of such magnitude that it not only vapourised the impacting body but also a substantial portion of the Earth's outer layers. The debris, consisting of material from both the Earth and the impactor, formed a ring around the planet. Over time, this material coalesced under the force of gravity to form the Moon, a process that likely took less than a hundred years to commence but may have taken tens of millions of years to complete. It's hard to comprehend how debris from an explosion could come together and form such a perfect shape as our moon!

The formation of the moon through this giant impact hypothesis explains several of its physical characteristics and the peculiarities of its orbit. For instance, the moon's composition closely mirrors that of the Earth's outer layers, lacking in heavier, metallic elements that would be expected if the moon had formed elsewhere in the solar system and been captured by Earth's gravity. This similarity supports the theory that the moon is essentially made of the same material that constitutes the Earth's mantle and crust.

The impact theory accounts for the current dynamics of the moon's orbit and its influence on Earth, including the stabilisation of the Earth's tilt and the relatively stable climate over geological timescales. The moon's gravitational influence on Earth has been a crucial factor in the development of life, affecting the tides and helping to regulate the planet's climate.

The energy imparted by the impact had other lasting effects on the structure of the moon and geological evolution. Initially, the moon existed in a molten state, with a magma ocean covering its surface. As it cooled, the lunar crust formed, primarily composed of lighter minerals that floated to the surface, while the heavier elements sank, creating a differentiated body with distinct layers. This early differentiation is a crucial process that has shaped not only the composition of the moon but also its geological features.

Over billions of years, the moon's surface has been subjected to intense bombardment by meteorites and comets, a history etched into its craters and basins. The maria, large basaltic plains, are remnants of ancient volcanic activity triggered by the heat from decaying radioactive elements and the tidal forces exerted by the Earth. These features, along with the highlands, tell a story of a dynamic world, shaped by internal processes and external forces.

Analysis of lunar samples has provided direct evidence supporting the theory, from the age of the moon to the similarities in isotopic compositions between Earth and lunar materials.

The formation of the moon is a story of destruction and creation, a testament to the dynamic processes that have shaped our solar system.

The early state of the moon

Following its tumultuous formation, the moon found itself in a fervently volatile state, undergoing rapid transformations. This initial phase of the moon's existence was dominated by an extensive magma ocean that enveloped its surface, a direct consequence of the immense heat generated by the impact that led to its creation, as well as the subsequent accumulation and compression of debris. The heat was so intense that it melted the outer layers of the nascent moon, creating a global ocean of molten rock that spanned several hundred kilometres in depth.

In this molten state, the moon began the slow process of differentiation, a critical phase that would dictate its geological future. Differentiation in planetary science refers to the gravitational separation of a planet's or moon's materials, based on their density. Heavier elements sink to form the core, while lighter materials rise to the surface, leading to the development of distinct layers. On the moon, this process was instrumental in the formation of its crust and mantle. As the magma ocean began to cool, minerals started to crystallise and segregate. The first minerals to crystallise were those rich in calcium and aluminium, which, being less dense than the molten magma, floated to the surface, forming the primordial lunar crust.

Simultaneously, denser materials, including those rich in magnesium and iron, sank, contributing to the formation of the lunar mantle. This period of cooling and differentiation was not swift; it is estimated to have taken tens of millions of years for the magma ocean to fully solidify and for the moon to develop a solid crust overlaying its mantle. The crystallisation of the magma ocean left behind a primordial crust predominantly composed of anorthosite, a rock characterised by a high content of plagioclase feldspar, which is reflective and gives the moon its bright appearance in the night sky.

This early state of the moon was crucial not only for its internal structure but also for setting the stage for its later geological activity and the appearance of its surface. The thickness of the crust, formed from the floating anorthosite, varied across the moon, being thinner where the maria would later form. These areas were susceptible to volcanic activity, as the thinner crust could more easily be breached by volcanic eruptions, leading to the flooding of these regions with basaltic lava, creating the dark plains visible from Earth.

In addition, the process of differentiation and cooling had implications for the moon's tectonic activity. Although the moon lacks plate tectonics that exist on Earth, its early cooling resulted in the contraction and fracturing of the lunar surface, creating tectonic features with unusual names—such as rilles and grabens. These features confirms the moon's dynamic early history and the forces that shaped its evolution from a molten state to the solid, differentiated body we observe today.

Cratered appearance

As the newly formed moon transitioned from a molten state to a solidified celestial body, its surface became the canvas for a relentless barrage of cosmic events that sculpted its appearance into the familiar cratered façade observed today. This process commenced shortly after its formation and has continued over the span of billions of years, marking the moon not only as a satellite of the Earth but as a chronicle of the solar system's tumultuous history.

The surface of the moon, devoid of significant atmospheric protection, has been exposed to an incessant stream of meteorites, asteroids, and comets. These celestial objects, remnants of the early solar system, collide with the moon's surface at high velocities, releasing tremendous amounts of energy. The impact of these objects excavates vast amounts of lunar material, creating craters that range from small, bowl-shaped cavities to immense basins spanning hundreds of kilometres in diameter. The energy released by these impacts is so intense that it can melt and vaporise the rock at the point of impact, with the ejected material falling back to the surface to form a distinct blanket of debris around the crater, known as ejecta.

This cratering process is responsible for the rugged, pockmarked appearance of the moon. Unlike Earth, where erosional processes such as wind, water, and tectonic activity continuously reshape the surface, the moon lacks an atmosphere, hydrosphere, or active tectonics capable of significantly altering these impact craters over time. Consequently, the lunar surface serves as a preserved historical record, with craters acting as timestamps that provide insights into the frequency and scale of impacts throughout the solar system's history

The largest and most ancient of these impact features, the lunar basins, were formed during a period known as the 'late heavy bombardment', which occurred around 4.1 to 3.8 billion years ago. This era witnessed a spike in the rate of asteroid impacts, not only on the moon but throughout the inner solar system. The origins of late heavy bombardment remain a subject of scientific debate, with theories suggesting it may have been triggered by the migration of the gas giants, which disturbed the asteroid belt and flung debris inward. The basins formed during this time are immense, often filled with solidified lava that later flowed onto the surface through fractures created by the impacts, forming the lunar maria. These dark, basaltic plains contrast sharply with the brighter, heavily cratered highlands, offering visual evidence of the moon's dynamic geological history.

The distribution and size of craters on the moon have been instrumental in developing the technique of crater counting, a method used by planetary scientists to estimate the ages of surfaces on other planets and moons. By comparing the density of craters on different parts of the moon's surface, researchers can infer relative ages, with older surfaces exhibiting more craters. This method has been central in constructing a chronological framework for the solar system's history

In addition to providing insights into the past, the study of lunar craters has contributed to our understanding of impact processes, including crater formation, ejecta dispersal, and the thermal effects of impacts. These findings have implications not only for planetary geology but also for assessing the impact hazard on Earth.

Maria and highlands

The moon's dark areas, traditionally called 'maria'—Latin for 'seas'—owe their name to early astronomers who, observing the moon through their telescopes, believed these vast, dark plains were filled with water, much like Earth's seas and oceans. This naming convention dates back to the seventeenth century when telescopic observations of the moon became possible, allowing astronomers to discern more details on its surface. The astronomer Galileo Galilei was among the first to describe these features in detail, and the names he and his contemporaries bestowed reflected their terrestrial analogies and the imaginative interpretations of the time. Over centuries, as lunar exploration advanced, it became clear that these 'seas' are not bodies of water but rather large, basaltic plains formed by ancient volcanic eruptions. Despite this clarification, the poetic names remain. There is a romance to somewhere named 'The Sea of Tranquility'. It was actually the *Apollo 14* mission that found the first conclusive evidence of water on the moon, although this had nothing to do with the maria.

The maria and highlands of the moon represent two starkly contrasting landscapes that tell the tale of lunar geological diversity and its dynamic past. These features, with their distinct appearances and compositions, offer profound insights into the moon's volcanic activity, its internal structure, and the processes that have shaped its surface over billions of years.

The lunar maria are the result of ancient volcanic eruptions that occurred between three and four billion years ago, filling the impact basins created by massive collisions with meteorites and asteroids during the moon's early history. The maria are predominantly found on the near side of the moon, which faces Earth, leading to speculation about the asymmetry in the moon's crustal thickness and volcanic activity between the near and far sides. The basalt that composes the maria is richer in iron and magnesium than the materials found in the highlands, and it is this composition that gives the maria their darker coloration, creating a striking contrast against the brighter lunar highlands.

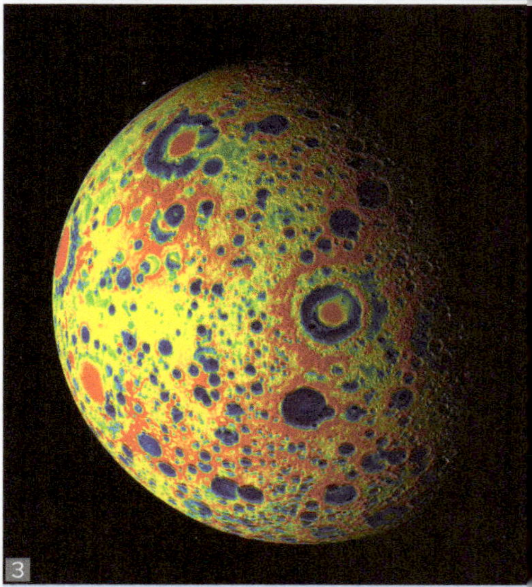

1. Apollo 11 astronauts checking the rocks they found on the moon. 2. Moon mountains. 3. The gravity mapping of the moon, showing its key points of gravitational pull. 4. Moons rocks and powder. 5. An artistic depiction of ancient astronomers observing the moon.

In contrast, the lunar highlands encompass the rugged, heavily cratered regions that occupy the majority of the moon's surface area. Composed primarily of anorthosite, a type of rock rich in calcium and aluminium silicates, the highlands are lighter in colour and represent the most ancient parts of the lunar surface, with some areas dating back over four billion years. This ancient crust, formed from the floating, cooling crust of the moon's magma ocean, has been subjected to relentless bombardment by celestial debris, giving it a rugged, mountainous appearance with peaks formed both by the accumulation of ejecta from impacts and by processes related to the impacts themselves.

The dichotomy between the maria and highlands is not merely superficial. It reflects the moon's geological complexity and the varied processes that have shaped its surface. The maria, with their smoother, younger surfaces, provide evidence of the moon's thermal evolution and the existence of a once-active internal geology that produced vast flows of basaltic lava. The presence of these plains amidst the highlands also illustrates the moon's asymmetrical geology, with a concentration of maria on the near side attributed to a combination of factors, including a thinner crust that allowed magma to reach the surface more readily.

The study of maria and highlands has been crucial for lunar exploration and science. Samples returned from the Apollo missions, primarily collected from the maria, have allowed scientists to directly date the volcanic flows and by extension, refine the chronology of the Moon's geological history. These samples have also provided insights into the Moon's interior, revealing details about its mantle composition and the thermal history that led to volcanic activity.

Moon dust

The regolith of the moon, something we may casually refer to as 'moon dust', is a layer of fragmented and loose material that blankets its surface. It has been created over billions of years. This fine-grained cloak, encompassing everything from powdery dust to broken rock fragments, is the product of the moon's exposure to the harsh environment of space, including the continuous bombardment by meteoroids and the effects of solar and cosmic radiation. But humans were to find out all about regolith first hand.

> *Sea of Tranquility 16 July 1969 23:24:19 hrs*
> **LMP (EVA) Lunar Module Pilot (Extravehicular Activity):** *Neil is now unveiling the plaque *** gear.*
> *[***Three asterisks denote clipping of words and phrases]*
> **CC Mission Control Centre:** *Roger. We got you boresighted, but back under one track.*
> **CDR Central Data Recording (EVA):** *For those who haven't read the plaque, we'll read the plaque that's on the front landing gear of this LM. First, there's two hemispheres, one showing each of the two hemispheres of the Earth. Underneath it says "Here, Man from the planet Earth first set foot upon the Moon, July 1969 A.D. We came in peace for all mankind." It has the crew members' signatures and the signature of the President of the United States.*
> **CDR (EVA):** *Ready for the camera?*
> **LMP (EVA):** *No. I'll get it. No, you take this TV on out.*
> **CDR (EVA):** *Watch the LEC, there.*
> **LMP (EVA):** *Now I'm afraid these ... materials are going to get dusty *** .*

> **LMP (EVA):** *The surface material is powdery. *** How good your lens is, but if you can *** smudges ... very much like a very finely powdered carbon, but it's very pretty looking.*

Moon dust was not as harmless as it looked. Once the astronauts were exposed, it got into everything back in the lunar module, despite leaving their moon boots outside. It cut like glass, smelled like gunpowder and probably wasn't that good for them.

Moon dust plays a crucial role in defining the moon's appearance and its thermal properties. The creation of this layer is primarily attributed to the impacts of meteoroids, ranging from microscopic dust grains to larger asteroids that strike the moon's surface with enough force to pulverise rock and soil. Over time, these impacts have not only excavated and redistributed material across the lunar surface but have also generated a layer of debris that varies in thickness from a few meters to tens of metres deep. This variation in depth is influenced by factors such as the prevalence of impact craters, with thicker deposits found in the older, more heavily cratered highlands and thinner layers in the younger maria.

The process of regolith formation involves not only mechanical fragmentation from impacts but also the alteration of rock by solar wind irradiation. The solar wind, a stream of charged particles emanating from the Sun, bombards the lunar surface, implanting atoms into the upper layers of the regolith and causing chemical changes in the rocks and dust. This interaction has significant implications for the study of the lunar surface, as the regolith acts as a repository for solar wind particles, including helium-3, a potential resource for future lunar explorers and a candidate fuel for fusion power.

The physical properties of lunar regolith are markedly different from soils on Earth. Due to the lack of weathering processes like wind and water erosion, the particles of lunar regolith are sharp and angular, a characteristic that poses challenges for both human activities and mechanical operations on the moon's surface.

Despite these challenges, the regolith holds invaluable scientific information. Analysis of regolith samples returned by the Apollo and Luna missions has provided insights into the moon's composition, the history of solar activity, and the processes that shape the lunar surface. For instance, the isotopic signatures found in regolith samples have helped scientists understand the timing and duration of lunar volcanism and impact events.

The study of lunar regolith also has practical implications for future lunar exploration and potential habitation. Understanding its composition and properties is essential for the development of technologies for surviving on the moon. The ability to extract oxygen, water, and other useful materials from the regolith could significantly reduce the need to transport resources from Earth, making sustained lunar exploration and eventual colonisation more feasible.

Volcanic activity

It might seem surprising now, but there is good evidence of past volcanic activity on the moon. This is primarily visible through the extensive maria and distinctive volcanic domes. This volcanic activity, which played a crucial role in shaping the lunar surface, offers insights into the moon's internal dynamics and thermal evolution during its formative years.

The maria are the most visible reminder of the moon's volcanic past and their origin can be traced back to a time when the moon's interior was still hot enough to generate volcanic eruptions that could breach the surface. The magma, rising from the mantle through fractures created by massive impacts, spread out over large areas, cooling to form the basaltic plains we observe today.

In addition to the maria, the lunar surface is dotted with volcanic domes, smaller but significant features that further attest to the moon's volcanic history. These domes, which range in diameter from less than a kilometre to more than 20 kilometres, are believed to be the remnants of less viscous, or more fluid, basaltic lava eruptions. Unlike the vast plains of the maria, these domes represent localised volcanic activity, where magma was able to reach the surface through vents or fissures, forming shield volcanoes or lava plateaus. The domes' smooth, rounded shapes, and the presence of summit craters on some, are characteristic of volcanic formations seen on Earth and other planetary bodies, indicating a shared process of volcanic activity.

The study of lunar volcanic activity is not limited to surface observations. Analysis of lunar rocks returned by the Apollo missions has provided direct evidence of the Moon's volcanic past. These samples have allowed scientists to date the volcanic events, with some evidence suggesting localised activity may have continued until as recently as one billion years ago. This timeline offers valuable insights into the thermal history of the moon, suggesting that it remained geologically active for a significant portion of its history.

By comparing lunar volcanic features with those on Earth, Venus, Mars, and even some of the moons of the outer planets, scientists can gain insights into the similarities and differences in volcanic activity across the solar system. The basaltic plains of the maria, with their relatively smooth terrain, offer potential landing sites for future missions, while the volcanic rocks themselves may provide resources for construction, shielding, or other in-situ resource utilisation efforts.

Tectonic shifts

The moon's geological narrative is not solely one of volcanic activity and impact cratering; it also includes a chapter on tectonic shifts that have significantly contributed to shaping its surface. Despite the absence of active plate tectonics, the moon has experienced its form of tectonic activity, evident in features like lobate scarps and wrinkle ridges. These structures tell a story of a celestial body that has cooled and contracted over time, its crust responding to internal and external forces long after its volcanic fires dimmed, and the last major impacts had scarred its surface.

Lobate scarps are among the most telling of these tectonic features. They are essentially cliffs or thrust faults that formed as the moon's interior cooled and contracted, causing the brittle crust to break and one section to be thrust over another. These scarps can stretch for several kilometres in length and stand tens of metres high, marking the boundaries where the lunar crust has been compressed. Their distribution across the lunar surface, observed in both high-resolution images from lunar orbiters and in situ by the Apollo missions, indicates that this process of contraction is a global phenomenon, affecting much of the moon. Interestingly, some lobate scarps are relatively small and appear to crosscut younger geological features, suggesting that the moon has continued to cool and contract even in geologically recent times.

Wrinkle ridges, another significant tectonic feature, are found primarily within the lunar maria. Unlike the lobate scarps, which are evidence of compression, wrinkle ridges are thought to result from the cooling and contraction of the basaltic lava that fills the maria. As these vast plains of solidified lava cooled, they contracted, causing the surface to buckle and form ridges. These features can be kilometres long and are often complex in structure, with a central ridge flanked by smaller furrows and ridges. Their presence within the maria provides insight into the thermal evolution of these volcanic plains after their formation.

The study of lunar tectonics offers valuable comparisons with tectonic processes on other celestial bodies, including Mercury, where lobate scarps similar to those on the moon have been observed, suggesting similar contractional processes. Such comparisons enhance our understanding of the thermal evolution of rocky bodies in the solar system and the role of tectonic activity in shaping their surfaces.

Apollo missions

The Apollo missions marked an unparalleled epoch in lunar exploration and the collection of lunar rocks stands out as a cornerstone that has provided profound insights into the moon's geology. The samples returned to Earth by the Apollo astronauts have been instrumental in elucidating the moon's composition, its formation history, and the myriad processes that have sculpted its surface over billions of years.

These samples were meticulously gathered from a variety of locations across the moon's surface, including the vast lunar maria and the ancient highlands, providing a representative collection of the moon's diverse geological features.

The analysis of these lunar samples has yielded invaluable scientific dividends. For instance, radiometric dating of the rocks has provided definitive evidence of the moon's age—

about 4.5 billion years old—closely aligning with the age of the Earth and lending support to theories regarding their shared origin. The similarities and differences in the isotopic compositions of lunar and terrestrial rocks have been critical in refining models of the moon's formation, particularly the giant impact hypothesis.

In addition to volcanic rocks, the Apollo missions also returned samples of the moon's ancient crust, primarily composed of anorthosite. The analysis of these rocks has provided evidence of the early differentiation of the lunar magma ocean, a process that led to the formation of the crust and mantle. This differentiation is a fundamental aspect of planetary formation and evolution, offering clues to the early thermal state of the moon and the dynamics of its interior.

The lunar samples have also been central in understanding the process of impact cratering, not just on the moon but across the solar system. The highlands samples, bearing the scars of countless impacts, have helped to chronicle the history of meteoritic bombardment in the inner solar system, and this has implications for understanding the conditions on the early Earth and the environment in which life first emerged.

How old is the solar system?

The study of lunar rocks has significantly contributed to our understanding of not only the moon but also the broader narrative of solar system formation. The moon's geological record, encapsulated within these rocks, serves as a crucial archive of the early solar system, providing evidence of the conditions and processes that shaped its formation and evolution.

One of the fundamental insights gained from lunar samples is the timing of the solar system's formation of approximately 4.5

billion years, as confirmed by radiometric dating of lunar rocks, which means that by extension, this is also the approximate age of the solar system. This precise dating provides a temporal framework within which scientists can place the sequence of events that led to the formation of the solar system, from the collapse of a molecular cloud to the accretion of planets.

The lunar geological record has played a key role in understanding the process of planetary differentiation, a fundamental aspect of planet formation. The differentiation of the moon, as inferred from the diversity of rock types and their compositions, mirrors the broader process of differentiation that occurred in the early solar system, leading to the formation of planets with distinct layers (core, mantle, and crust). This process, driven by the heat from accretional impacts, radioactive decay, and gravitational compression, is critical for understanding the internal structure and composition of planets.

The isotopic compositions of lunar materials have offered clues to the provenance of the solar system's building blocks. The similarities and differences in isotopic ratios between lunar, terrestrial, and meteoritic materials have helped scientists trace the distribution of elements and isotopes in the early solar nebula, offering clues to the processes of solar and planetary formation.

Traces of water found in lunar rocks challenge earlier assumptions about the dry nature of the moon and, by extension, the conditions under which terrestrial planets formed. These findings suggest a more complex scenario of water distribution and retention during the early stages of planetary formation, with implications for the habitability of planets in the solar system and beyond.

The moon's interactions with Earth, from tidal forces to the stabilisation of Earth's axial tilt, underscore the importance of gravitational interactions between celestial bodies. These interactions have significant implications for the climatic stability of planets and the conditions necessary for sustaining life. It is quite likely that without the moon—we may not exist. But we do! And as we continue to explore the Moon, through both robotic missions and potential human exploration, we can expect to uncover further insights that will deepen our understanding of the solar system and the universe at large.

CHAPTER 5

Illuminating the divine: Moon worship across cultures

Today, many people only see the moon as a cutaway in a movie. In films, the moon can indicate outdoors, time passing, something supernatural, a nice moment in a nature documentary or nothing much at all. But before we spent our lives looking at screens, people had a big 3D version on display every night. It was their only show. It's rare for a planet to have a moon as big and close as we do, and it's even rarer to have intelligent life trying to decipher its meaning. While it's impossible to get into the actual minds of our ancestors, we'll try and gain some insights. If anything was worthy of worship, it was the big glowing thing in the night sky. It even seemed to have a face on it. We still worship the moon in our own way. Who doesn't like a warm night with a full moon in view? It makes you feel all mystical and sometimes you feel you could almost touch it.

Since the beginning of Earth time, or close to it (and putting aside the great big smashings that occured), this big orb has

shone on the planet, beaming down on molecules, cells, reptiles, beasts and stirring the hearts and minds of humanity. It is the unavoidable celestial companion of our planet, waxing and waning through the night sky, marking the passage of time and seasons. Across eras and continents, people have gazed up at the moon and seen in it the face of the divine. The history of moon worship is as old as humanity itself, a testament to our enduring fascination with the night sky's most luminous body.

Throughout the history of human culture, the moon has been a constant source of inspiration. Its influence is woven into the fabric of our myths, our art, and our religious practices. To our ancestors, the moon was more than just a celestial body; it was a deity, a potent force of nature, and a beacon in the darkness, guiding the rhythms of life. From the tides that sweep our shores to the cycles that govern our bodies, the moon's subtle pull is omnipresent, and its worship is a common thread that links disparate cultures.

The allure of the moon transcends geographical boundaries and historical epochs. It has shaped the narratives of ancient civilisations and sparked the imaginations of poets and philosophers. Its phases have served as a celestial clock and calendar, its eclipses have been omens of change, and its steadfast presence has been a comfort to those who travel by night.

So look up and wonder, as we explore the rich and varied traditions of moon worship. From the temple ruins of the ancient Near East to the druggy modern festivals, we will uncover the stories of lunar gods and goddesses, the rituals that honour the moon's power, and the enduring legacy of this ancient practice. We will delve into the ways in which moon worship has evolved over the millennia, adapting to the changing tides of human belief and remaining a compelling force in the spiritual life of countless individuals. The moon's story is our story, a reflection of humanity's quest for meaning among the stars. Let us now illuminate the divine paths carved by the moon's soft radiance, as we traverse the history of its worship and its sacred place in the human soul.

The luminous god: Early human cultures and lunar reverence

Today, we can only speculate about the experiences of early human cultures. It's hard not to get overly mystical about their moon experiences. We can only imagine. In the infancy of civilisation, under the vast, star-studded sky, early humans looked to the moon with a sense of wonder and, perhaps, fear. This great object in the sky, unlike the sun, did not command the day with a blazing force but rather commanded the night with a gentle authority that guided the earliest of humankind through the darkness. The moon was not just a nocturnal lantern; it was a celestial deity, a luminous presence that held sway over the night and all that happened beneath its watchful gaze.

The reverence of the moon in early human cultures often stemmed from its mysterious nature. The moon existed in a realm of dreams and shadows, a time when the veil between the mundane and the mystical seemed thinnest. Its changing face—the cyclical transformation from new to full—was a phenomenon devoid of the human hand, a natural magic that was both reliable and unfathomable. To our ancestors, these lunar phases mirrored life's own rhythms: birth, growth, decline, and rebirth.

Archaeological evidence, such as the alignment of ancient structures to lunar cycles and the discovery of artifacts bearing

ILLUMINATING THE DIVINE: MOON WORSHIP ACROSS CULTURES

the moon's image, suggests that lunar worship was widespread. Stonehenge (3000 to 1520 BCE), the famous prehistoric monument in England, is believed by some scholars to have been used to predict lunar eclipses, demonstrating the significance of the moon in ritual and timekeeping. Similarly, cave paintings from the Upper Palaeolithic period often depict the moon and its phases, hinting at a lunar calendar used to track time.

In these nascent societies, the moon also assumed a vital role in mythology and religion. Various cultures personified the moon in a myriad of forms. Some saw it as a powerful god or a fierce goddess, while others interpreted it as a symbol of fertility and rejuvenation. The moon's influence extended to the realm of agriculture, where its cycles were linked to the timing of planting and harvesting, and to the domain of fertility, where its waxing and waning were thought to reflect the human reproductive cycle.

The moon's omnipresence in the night sky made it a natural object for early religious focus. It became the subject of rituals and ceremonies—many of which have left no record beyond the alignment of stones or the remnants of ancient fires. Moon worship at this stage was not simply an appreciation of beauty or a scientific curiosity; it was a fundamental aspect of cultural identity and survival. It shaped calendars, guided nocturnal navigation, and even played a part in the governance of early societies, with leaders often seen as embodiments or representatives of the moon's divine will.

As we move through history, the moon's role in human culture becomes more defined, shaping deities and myths that would stand the test of time. Yet, this early lunar reverence laid the groundwork for a relationship with the moon that remains enigmatic and profound. It reminds us that our connection with the lunar sphere is as old as humanity itself—a connection that speaks of awe, fear, and the quest for understanding within the human spirit.

The cradle of civilisation: Moon worship in the Ancient Near East

The fertile crescent of the Ancient Near East, emerged as a cradle of early civilisation, from as far back as 12,000 BCE. Here, moon worship found sophisticated expression in the city-states of Mesopotamia. The lunar deity was not just an aspect of the divine but a central figure in the pantheon, commanding temples, priesthoods, and elaborate rites.

In Sumer, the world's oldest known civilisation, spanning from the fifth to third millennia BCE, the moon was deified as Nanna, a god of immense power and prestige. Nanna's worship was a complex blend of astronomy and theology, with priests charting the moon's course to divine omens and direct earthly affairs.

The ziggurats, towering temple structures, served not only as places of worship but also as observatories from which the priesthood could study the moon and stars. Nanna's influence was such that worship persisted through successive empires, with later civilisations like the Akkadians and Babylonians adapting this god into their own religious frameworks as Sin, a name echoing through history as a testament to the moon's enduring divinity.

In these ancient societies, the lunar deity was often associated with justice and wisdom, attributes reflecting the moon's role in illuminating the night. Sin's temple in the city of Ur, one of the most important religious centres, was a hub of both worship and social order. The moon was a ruler of the sky and a keeper of knowledge, its cycles used to enact laws and dispense judgments.

The moon's influence extended beyond the temple walls and into the cultural and literary life of the Ancient Near East. Lunar deities featured prominently in myths and epics, with the moon often portrayed as a mediator between the forces of chaos and order.

In the Epic of Gilgamesh, one of humanity's oldest literary works, the hero Gilgamesh seeks out Utnapishtim, a figure granted immortality by the gods after surviving a great flood, an event marked by celestial alignments including the moon.

The reverence for the lunar deity in the Ancient Near East was also reflected in personal devotion. Amulets, seals, and small altars dedicated to the moon god have been unearthed, suggesting that worship was not confined to the priesthood but was a part of everyday life. The moon's phases were woven into personal prayers and public festivals, highlighting a society deeply attuned to the celestial rhythm.

The lunar deity's significance was often linked to fertility and prosperity. The waxing and waning of the moon became metaphors for the ebb and flow of resources and fortunes. Sin's benevolence was invoked for bountiful harvests and the protection of livestock, crucial aspects of survival in these agrarian societies.

As the heart of the Ancient Near East, the moon's divinity cast a long shadow, shaping the spirituality and daily practices of its people. The legacy of lunar worship in this region set the stage for future civilisations, influencing religious thought and practice across the Middle East and beyond. The celestial had become terrestrial.

The moon in Ancient Egypt

In Ancient Egypt, the moon was a divine figure of night and renewal. In the pantheon of gods, Khonsu, the god of the moon, reigned with a power that waxed and waned with the lunar cycle, at the very foundations of Egyptian culture.

The grandeur of Khonsu's worship is perhaps best exemplified by the monumental Karnak Temple Complex in Thebes, which reached its zenith around 1570 BCE. Here, within the vast precincts dedicated to the Theban triad, Khonsu was honoured as a deity of rejuvenation, with temples standing as a stone-bound chronicle of the moon's eternal dance through the heavens.

In Egyptian mythology, the moon was a symbol intertwined with the narrative of life, death, and the promise of rebirth. The enduring myth of Osiris, god of the afterlife, whose life cycle aligned with the moon's phases, resonated deeply with the Egyptian understanding of immortality. The new moon, akin to the dismembered and reassembling Osiris, spoke of renewal, the waxing moon of burgeoning life, and the full moon of complete rebirth, a cycle that played out endlessly across the millennia.

The celestial bodies held sway over the Egyptian cosmological belief system, with the sun and moon serving as the dual eyes of Horus, the falcon-headed sky god. The moon represented the left eye, reflecting the world of the gods and the mysteries of the divine. Its regular rebirth from the darkness of the new moon was a cosmic affirmation of the pharaoh's own divine regeneration and the continuity of his reign.

The lunar calendar, which dominated Egyptian timekeeping until the shift toward a solar calendar in the early dynastic period, was integral to religious festivities, agricultural routines, and the grand architectural endeavours that defined its skyline.

Mythological lore further wove the moon into tales of gods journeying through the underworld, with lunar eclipses casting dramatic shadows that echoed these mythical tales. Each emergence of the moon from eclipse was a retelling of triumph over chaos, a celestial game enacted upon the grand stage of the sky, reinforcing the order of the universe.

Inscriptions and papyri, such as the famous Edwin Smith Papyrus, believed to be written c. 1600 BCE, detail the moon's phases in conjunction with medical practices, showcasing the unique Egyptian conflation of the lunar cycle with the art of healing. This connection between the moon and wellbeing was emblematic of a civilisation that sought to understand the mysteries of life through the patterns of the heavens.

The moon's journey across the Egyptian night sky was not merely a passage of light; it was a narrative steeped in divinity, a tale that extended from the pre-dynastic era through the Ptolemaic period, concluding with the end of pharaonic rule around 30 BCE.

In the vast expanse of desert sands and along the Nile's fertile crescent, the lunar deity's role transcended the ages, serving as a guardian of time, a healer of the sick, and an emblem of the cycle of life and afterlife. Khonsu's luminescence, a glint of immortality, cast a silver line across the epochs,

The moon becomes female: Greco–Roman mythology

In Greece, as early as the Homeric poems of the eighth century BCE, the moon was personified by Artemis, the virginal huntress and twin sister of Apollo, the sun god. Artemis was not only the protector of the wilderness and the patroness of

1. Super blue blood moon.
2. Ancient Egyptian moon god. 3. Khonsu, another ancient Egyptian moon god. 4. Illustration of Greek moon workship, c. 800 BCE. 5. An ancient sunrise seen from the moon. Same as today!

hunters but also the guardian of childbirth and the young, reflecting the moon's connection to fertility and the natural cycles of life. Her worship was widespread and varied, with festivals such as the Brauronia and the Munichia held in her honour, where the luminosity of the moon was celebrated with torchlit processions and sacred rites. And here these ancient gods spill into modern life.

As with most Greek mythology, these twins are doing similar things but fifty years apart. The world became familiar with the sun god Apollo during the moon landings in the 1960s and 1970s and Artemis is the name bestowed on the current return to the moon program. *Artemis I*, an uncrewed Orion capsule has already successfully orbited the moon, in 2022, followed by *Artemis II*, a crewed mission.

Advancing through the centuries, the Hellenistic period, which began with the conquests of Alexander the Great in the fourth century BCE, saw the continuation and expansion of lunar veneration. Artemis' temples dotted the Mediterranean, and her cult adapted to local customs, further entrenching the moon's spiritual presence across the Hellenistic world.

The Romans, inheriting the Greek affinity for lunar deities, venerated the moon under the guise of Diana, the goddess of the hunt and the moon, blending her worship with their Italic traditions. By the first century BCE, Diana's sanctuary at Lake Nemi, known as 'Diana's Mirror,' became the centre of her worship, where the moon's reflection on the water's surface was seen as an act of divine communication. The annual festival of Nemoralia, held on the Ides of August, saw devotees lighting candles and torches in honour of the moon, their flames echoing its serene light.

The Roman calendar was replete with observances tied to the lunar cycle, reflecting the moon's temporal influence. The Ides of each month, falling on the first full moon, were days of religious observation and reflection. The Julian calendar reform of 45 BCE, instituted by Julius Caesar, was a turning point in timekeeping, as it attempted to align the calendar year with the solar year rather than the lunar cycle, showcasing a shift in the cultural approach to celestial bodies.

In Greco–Roman mythology, the moon's phases played a symbolic role in the tales of deities and heroes, often representing purity, change, and the passage of time. The poets of antiquity, from Hesiod in the seventh century BCE to Ovid in the first century BCE, sang of the moon's beauty and mystery, embedding it in the cultural consciousness as a source of poetic inspiration and philosophical contemplation.

> ***The Moon by Sapho (circa 590 BC; English Translation, 1893)***
> *The stars about the lovely moon*
> *Fade back and vanish very soon,*
> *When, round and full, her silver face*
> *Swims into sight, and lights all space.*

Throughout the Greco–Roman world, the moon was a celestial muse, a character in a grand narrative that spanned from the lyric verses of Sappho to the epic poetry of Virgil. Its presence in literature and art from the Archaic period through the twilight of the Roman Empire remained a constant, a reminder of the eternal interplay between the earthly and the divine.

As the Western Roman Empire waned, the reverence for the lunar deities transitioned, merging with the emerging Christian ethos, yet the classical moon's legacy persisted. The moon, throughout the Greco–Roman epoch, was a symbol of harmony and reflection, a guiding light for civilisations that sought to find balance between the rational and the mystical, the seen and the unseen, the temporal and the eternal.

The Eastern luminary: Lunar deities of Asia

In the ancient civilisations of China, the moon was personified and deified in the form of Chang'e, the moon goddess whose tale became a cornerstone of Chinese mythology and folklore. Again her name has been appropriated for the Chinese space program. Her legend, dating back to the Warring States period (c. 481–221 BCE), is celebrated during the Mid-Autumn Festival, a tradition that has endured for millennia. This festival, held on the fifteenth day of the eighth lunar month, sees families gather to honour the full moon and its symbol of completeness and unity. Mooncakes, a traditional delicacy, are shared, emulating the moon's shape and embodying the spirit of reunion.

Further to the south, in the Indian subcontinent, the moon's divine aspect took shape in the form of Chandra or Soma, the lunar deity of Hinduism. The Rigveda, one of the oldest existing texts in any Indo–European language, composed around 1500 BCE, reveres the moon as a harbinger of nourishment and soma, the elixir of immortality. Chandra's significance weaves through the fabric of Hindu culture, manifesting in the calendar where days are named after the phases of the moon, and festivals such as Karva Chauth and Purnima are celebrated in its honour.

In the Shinto and Buddhist traditions of Japan, the moon deity Tsukuyomi-no-Mikoto, or simply Tsukuyomi, emerges from the ancient myths of the Nihon Shoki and the Kojiki, texts compiled in the eighth century CE. Tsukuyomi, born from the right eye of the primordial deity Izanagi, was celebrated for his beauty and serenity. The Otsukimi, or 'Moon-Viewing' festival, established during the Heian period (794–1185 CE), is a cultural embodiment of lunar appreciation, where the Japanese people observe the harvest moon, offering rice dumplings as a token of gratitude for the year's harvest.

Throughout the vast continent of Asia, the moon's worship is not confined to the mythological and the celebratory; it also permeates the astrological and divinatory practices ingrained in these cultures. The Chinese lunar calendar, a system that dates back to the Shang dynasty (c.1600–1046 BCE), continues to influence festivals, agricultural practices, and even personal decisions in life.

The moon's reflection in Indigenous traditions

There is a wide variety of ways the moon is perceived in the many cultures that are now labelled Indigenous. From these sources, the moon emerges not just as a celestial body but as a profound symbol of connection to the natural world, spirituality, and the cycles of life and death. These traditions, deeply rooted in the landscapes from which they arose, offer a diverse spectrum of lunar reverence that speaks to the universal human experience of finding meaning in the heavens.

Among the Indigenous peoples of North America, the moon has long been a central figure in storytelling, ceremony, and the marking of time. The Indigenous American tribes, with their rich oral traditions and deep spiritual connection to the land, have myriad moon myths that vary greatly from one nation to another. The Haudenosaunee (Iroquois) Confederacy, for example, recognises the moon as a bringer of light in the darkness, with ceremonies such as the Midwinter Ceremony, which aligns with the lunar calendar to renew the year and the spirit of the people.

In Australia, diverse Indigenous cultures, which have thrived on the continent for over 50,000 years, view the moon as an integral part of their story, the foundational mythology that explains the origins and culture of the land and its people. The moon's phases are intertwined with tales of creation, transformation, and the cyclical nature of existence, reflecting a deep understanding of the environment and the interconnections of life.

The influence in the transition of beliefs: polytheism to monotheism

As ancient Middle Eastern civilisations morphed, religious beliefs shifted from the many gods of polytheism to the singular deity of monotheism. The moon's place in the spiritual consciousness of these people changed. But it still hung in there, subtly shining on the fabric of monotheistic religions and continuing to illuminate the spiritual landscape of billions. In the traditions of Judaism, Christianity, and Islam—religions that have shaped the course of recent history and the destiny of many nations—the moon transitioned from a divine entity to a symbol and signifier of time, festivity, and reflection.

Judaism (c. 1800 BCE – present): In Judaism, the lunar cycle plays a pivotal role in the Hebrew calendar, a lunisolar system that determines the dates of Sabbath, festivals, and holy days. The new moon, or Rosh Chodesh, marks the beginning of each month and is a time for renewal and celebration. The sighting of the crescent moon once heralded communal observances and offerings, linking the celestial to the covenant between god and Israel. The festivals of Passover and Sukkot, among others, are timed according to the lunar calendar, ensuring that these ancient commemorations align with the seasons and agricultural cycles as commanded in the Torah.

Christianity (first century CE – present): Christianity, while primarily solar in its observance of time, retains the lunar connection in determining the date of Easter, the most significant Christian festival celebrating the resurrection of Jesus Christ. The Council of Nicaea in 325 CE established that Easter would be observed on the first Sunday after the first full moon occurring on or after the vernal equinox. This decision weaves the moon into the fabric of Christian worship, linking the lunar cycle to the foundational event of Christian faith.

Islam (seventh century CE – present): Islam embraces the moon with profound reverence, utilising a purely lunar calendar to determine the dates of Ramadan, Eid al-Fitr, Eid al-Adha, and other Islamic observances. The crescent moon, sighted to mark the beginning and end of Ramadan, is a powerful symbol of Islam, representing the passage of time, the rhythm of the faithful's life, and the cycle of fasting and feasting. The Hijra, or the migration of the Prophet Muhammad from Mecca to Medina, which marks the beginning of the Islamic calendar, underscores the moon's significance in marking a new era of religious and communal identity.

Pagan traditions: The revival of lunar worship: In contemporary times, the revival of pagan traditions has brought with it a renewed interest in the moon as a symbol of feminine power, fertility, and magic. Wicca, a modern pagan religion that emerged in the mid-twentieth century, places the moon at the heart of its worship, with rituals and spells often timed to the phases of the moon. The Esbat, a ceremony conducted at the full moon, celebrates the goddess's aspect as the moon, embracing her energy for healing, growth, and protection.

Similarly, in other neopagan traditions, such as Druidism and various forms of witchcraft, the moon's cycles guide celebrations, rituals, and the practice of magic. The new moon and full moon serve as times for reflection, renewal, and manifestation, embodying the eternal cycle of death and rebirth that is central to pagan cosmology. We will return to these current moon fashions in later chapters. The moon is still the go-to orb for any group attempting to inject a little something special into their cult. But we will end this chapter with a word from the ancient Roman poet Ovid: imagining a time before the moon!

> **'Metamorphoses' by Ovid, 8 CE**
> **Book I**
> *My mind is bent to tell of bodies changed into new forms. Ye gods, for you yourselves have wrought the changes, breathe on these my undertakings, and bring down my song in unbroken strains from the world's very beginning even unto the present time.*
> *Before the sea was, and the lands, and the sky that hangs over all, the face of Nature showed alike in her whole round, which state have men called chaos: a rough, unordered mass of things, nothing at all save lifeless bulk and warring seeds of ill-matched elements heaped in one. No sun as yet shone forth upon the world, nor did the waxing moon renew her slender horns; not yet did the earth hang poised by her own weight in the circumambient air, nor had the ocean stretched her arms along the far reaches of the lands. And, though there was both land and sea and air, no one could tread that land, or swim that sea; and the air was dark. No form of things remained the same; all objects were at odds, for within one body cold things strove with hot, and moist with dry, soft things with hard, things having weight with weightless things.*

CHAPTER 6

Moon mythology, werewolves, vampires and romantics

The moon is not just a celestial charmer hanging in our night sky, a beacon for lovers and poets, but a stage light for darker dramas where myth and magic collide. Have you ever gazed upon a full moon and felt a shiver down your spine? You're not alone. The moon has been part of terrifying tales since forever, playing a starring role in the mythology of almost every culture under ...the moon.

First off, let's talk about the moon's critical status in the world of myths. It's like the cool, mysterious character in every fantasy saga that you can't help but be intrigued by. From lighting the path to the underworld, to being the ultimate plus one at supernatural gatherings, the moon knows how to keep things interesting. It's got connections that would make even the most seasoned Hollywood agent jealous, rubbing celestial elbows with underworld deities and all sorts of night-crawling critters.

As for night-time revelers—werewolves and vampires have been moonlighting as the moon's plus-ones for centuries.

These aren't your typical party animals, though. Under the moon's influence, they transform. It's like the moon has this magical touch, a supernatural remote control that can turn man to beast or give the undead a night out.

But it's not all fur and fangs. The moon also has this profound, mystical mojo that seems to touch everything in the natural and mystical world. It's like that friend who's into crystals and tarot readings, but way more powerful. Imagine being able to control the tides or mess with the natural order of things—that's the moon for you, a real influencer in the planetary world.

And then there's the whole transformation thing. The moon doesn't just sit pretty in the sky; it's an agent of change. Think of it as the ultimate life coach, guiding humans (and not-so-humans) through their personal glow-ups and meltdowns, often with a furry twist. Whether it's a werewolf howling at the sky or a mortal tapping into their inner beast, the moon's got a way of revealing the hidden behind the shadows.

And let's not forget the moon's dual personality: it's the original master of duality, playing both sides of the cosmic coin. On one hand, it's all gentle illumination and guiding light, and on the other, it's dark, mysterious, and a bit Goth. It knows how to keep the balance showing us that both sides have their moments in which to shine.

What's intriguing is how the moon plays on our fears and dreams. It's like that enigmatic character in stories that you're drawn to but slightly wary of. It holds a mirror to our fascination with the unknown, reminding us that there's so much more beyond our world, whispering tales of adventure and caution in the same breath.

And let's not brush past our furry friends, the werewolves, and the eternally chic vampires, each a testament to the moon's power to bridge worlds and stir up the eternal cocktail of life, death, and immortality. They're the moon's partners in crime, showcasing the eternal dance of light and shadow, and frankly, making eternity look pretty good.

So, the next time you find yourself moon-gazing, remember, you're not just looking at a rock in space. You're peeking into the heart of countless tales, a world where myth and magic reign, and the night is just a curtain waiting to be drawn. The moon, with its mysterious aura and legendary entourage, continues to spin tales that captivate the imagination, reminding us that in the dance of light and darkness, there's always a story to be told.

The moon as a mythological symbol

The moon, with its ethereal glow and cyclical journey across the night sky, has long captivated the human imagination, embedding itself as a prominent symbol in the tapestry of world mythologies. This celestial body, marked by its phases of waxing and waning, embodies a multitude of meanings and associations, ranging from the mysterious to the supernatural. Its consistent presence in the night sky, coupled with its ever-changing nature, offers a paradoxical blend of constancy and transformation that has inspired countless myths and legends across cultures. It was an obvious thing to build a story around. Unlike the sun, you can look at it for a long long time and wait for inspiration.

In the realm of mythology, the moon is not merely a physical entity orbiting the Earth but a potent symbol intertwined with the fabric of the stories that cultures tell about the origins of the world, the gods, and the cosmos. It often represents the feminine principle, linked to goddesses of fertility, magic, and the night. These lunar deities, found in cultures from the ancient Near East to the shores of the Pacific, oversee the realms of birth, death, and rebirth, echoing the moon's own cycles of renewal.

The moon's association with the supernatural extends to its influence on mythical creatures, whose lives and powers are often depicted as intertwined with lunar phases. The classic examples of werewolves and vampires, creatures of transformation and the night, illustrate how the moon's phases—particularly the full moon—have been believed to trigger metamorphoses and to enhance supernatural abilities. This connection underscores the moon's dual role as both a beacon in the darkness and a gateway to the unknown and the otherworldly.

The moon's link to the mysterious and the supernatural is also deeply embedded in the concept of the underworld in mythology. The underworld, a realm of death and the afterlife, is often portrayed as a place of shadows and secrets, where the moon acts as a guide for the souls of the dead or as a symbol of hope and resurrection. This association highlights the moon's role in the cycle of life and death, mirroring its own phases of disappearance and reappearance in the night sky.

The cyclical nature of the moon is a powerful metaphor for the cycles of nature and human life, embodying the themes of growth, decay, and renewal. Just as the moon undergoes its phases, so too do individuals and societies experience cycles of struggle, change, and rebirth. These universal patterns, reflected in the lunar cycle, have imbued the moon with a symbolic significance that transcends its physical presence, making it a central figure in mythological narratives.

The moon's ability to illuminate the night sky, revealing what is hidden in darkness, has also led to its association with enlightenment and knowledge. In many mythologies, the moon is seen as a source of wisdom, its light a metaphor for the pursuit of truth and understanding amidst the shadows of ignorance and fear. This aspect of the moon as a bringer of clarity and insight complements its more mysterious and enigmatic qualities, offering a balanced portrayal of its influence on the human psyche and the natural world.

Association with the underworld

In the vast expanse of human belief and mythology, the moon has played a common role as a bridge to the metaphysical realms that lie beyond our tangible existence. Among these realms, the underworld holds a special place, often envisioned as a domain where the finality of death intersects with the mystical. The moon's connection to this enigmatic realm extends far beyond mere symbolism, embodying a profound relationship that explores the complexities of life, death, and the human soul's journey.

The underworld, in various mythological traditions, is portrayed as a hidden layer of existence, a shadowy afterlife where souls venture after their earthly departure. This realm, while feared by many, is also a source of deep wisdom, transformation, and renewal. The moon, with its ethereal glow, serves as a guiding light for these souls, illuminating paths shrouded in darkness and mystery. Its cyclic disappearance and return from the night sky mirror the human fascination with the concepts of rebirth and continuity beyond physical demise, suggesting a cyclical nature of existence that parallels the moon's own phases.

As the moon waxes, it reflects the soul's journey through the underworld, facing trials and gaining insights. The full moon, then, symbolises enlightenment and completion, the soul's readiness to transition to a new phase of existence, whether it be a return to the living world or a transcendence to a higher state of consciousness.

The moon's influence extends to the deities and entities that govern the underworld. Many cultures revere moon gods or goddesses who possess dual roles as custodians of the night and guides for the deceased. These deities often embody qualities of both compassion and severity, reflecting the moon's own dual nature as a source of light and a harbinger of darkness. Through rituals and offerings to these lunar deities, ancient peoples sought protection for the deceased and assurance of safe passage through the underworld's trials.

The moon's connection with the underworld also manifests in the lore surrounding mythical creatures and spirits that are believed to roam the night. These beings, often seen as messengers between the worlds of the living and the dead, reinforce the moon's role as a mediator and conduit between different planes of existence. Their activities, heightened under the moon's light, serve as reminders of the thin veil that separates the mundane from the mystical, urging a deeper contemplation of life's mysteries and the nature of the afterlife.

Connection to mythical creatures

Among the moon's imagined animal world, werewolves and vampires stand out, their legends intricately linked with the lunar cycle, embodying the transformative and enigmatic power of the moon.

Werewolves, creatures of transformation, are perhaps the most iconic symbols of the moon's influence over myth and folklore. The transformation of a human into a werewolf under the full moon is a powerful image that captures the imagination, symbolising the untamed, primal forces that lie dormant within all of us. Basically, the full moon turns an ordinary looking human into a savage wolf. Developments in special effects in cinema have made it possible for us to look at this fairly unlikely transformation in various stages. It goes back to the ancient belief in the moon as a catalyst for change and revelation, exposing hidden truths and bringing forth the inner beast. The werewolf legend, in its essence, is a tale of duality and struggle, mirroring the moon's own dual nature as a symbol of light in darkness and a gateway to the unknown.

Vampires, on the other hand, embody a different aspect of the moon's mythical connections. Again filmic effects have illustrated their lifestyles over the last century. Beginning with *Nosferatu*, the German classic in 1922. These nocturnal creatures, often depicted as drawing strength from the moon, navigate the night with a grace and power that is both alluring and terrifying. The moonlight not only provides them with the cover of darkness to hunt but also symbolises the blurred lines between life and death, immortality and damnation. Vampires, like the moon, are entities of contrast and contradiction, existing on the fringes of the human world, evoking both fear and fascination. Their association with the moon enhances their mystical aura, casting them as eternal wanderers under its watchful eye, forever bound to the cycle of night and day.

The legends of werewolves and vampires, enriched by their connection to the moon, serve as metaphors for the human condition, exploring themes of transformation, immortality, and the shadow self. These mythical creatures, illuminated by the moon's light, invite us to ponder the mysteries of existence and our place in the universe. They remind us of the thin veil that separates the known from the unknown, the seen from the unseen, urging us to look beyond the surface and embrace the complexity of the world around us.

These stories offer a window into the collective psyche. They reflect our enduring interest in the boundaries of human experience and the possibilities that lie beyond the rational and the ordinary. In this way, the moon and its mythical creatures enrich our cultural imagination, providing stories that transcend time and space, inviting us to explore the depths of our fears, desires, and dreams. Werewolves and vampires howl and bite each other while we twitch in our sleep.

Werewolf v. vampire

That leads us to the question—who would win this mythical battle? It's all a human construct of course, brought to life in stories and movies. Or do these stories contain elements of truth? These nocturnal adversaries, each with their own unique strengths and vulnerabilities, have been locked in an eternal struggle for supremacy. At the heart of this epic conflict, the moon plays a pivotal role, tipping the scales of power in a dance as ancient as the creatures themselves.

Werewolves, creatures of raw strength and primal rage, draw their power from the moon. Under its glow, they transform into towering beasts of unmatched ferocity, their physical prowess magnified to terrifying proportions. With enhanced agility, superhuman strength, and a heightened sense of smell, werewolves become formidable hunters under the full moon. Their ability to heal rapidly from most wounds makes them nearly invincible, save for their Achilles' heel—silver.

Vampires, on the other hand, are creatures of cunning and supernatural might. Their strengths lie not in brute force but in their speed, their ability to regenerate, and their mastery of the dark arts. Immortal beings of the night, vampires possess an array of powers including superhuman strength, the ability to shape-shift, and control over lesser creatures. However, their reliance on human blood for sustenance and their vulnerability to sunlight are weaknesses that can be exploited.

Theoretically, in a battle under the full moon, where the werewolves draw the full extent of their power from the celestial body, they would have a significant advantage. Their increased strength and healing abilities, amplified by the lunar influence, would enable them to sustain and recover from a vampire's attacks, while their ferocity could overwhelm the more calculated and reserved vampires. However, the outcome of such a battle is not solely determined by physical prowess.

Vampires, with their strategic thinking and centuries of accumulated knowledge, could potentially outmaneuver the werewolves, using their weaknesses against them. The battle could turn on the vampires' ability to evade direct confrontation until the werewolves' lunar-induced advantage wanes with the setting moon.

Would the dumb dog lose? The outcome of this mythical battle is not clear-cut. It would ultimately depend on the circumstances of the encounter, the environment, and the ability of the vampires to strategically counter the werewolf's lunar-fueled ferocity. The battle between light and shadow, strength and strategy, is a delicate balance, and in the realm of myth, the victor of such a battle remains a tantalising question, shrouded in mystery and the whispers of the night. Wait for the movie of this book!

The belief in the moon's supernatural powers is rooted in its observable effects on the Earth and its inhabitants. The moon's influence on the natural world is evident and undeniable, from

1. Transylvanian vampires, as they would have appeared in the 1700s.
2. Dracula v. werewolf: who wins? 3. Werewolf during a full moon.
4. The orginal book cover of Bram Stoker's *Dracula*. 5. The horror story writing kit.
6. All kinds of boat tragedies happened on a full moon during the 19th century.
7. Frankenstein.

the control of tides to its impact on plant growth and animal behaviour. These tangible effects have served as a foundation for the belief in its more arcane powers, suggesting that if the moon holds sway over the seas and the earth, it must also affect the unseen aspects of existence. This leap from the observable to the mystical encapsulates the human desire to connect with the universe, to understand and harness the forces that govern life and destiny. And to tell stories.

Romantics, Gothics and the moon

The Romantic Movement, spanning the late eighteenth to the mid-nineteenth century, marked a period of profound artistic, literary, and intellectual expansion that was deeply intertwined with personal emotion and the sublime natural world. The movement was something of a reaction to an age where science and machines were transforming life, especially in Britain, the centre of the industrial revolution. Romanticism celebrated powers of nature, the glorification of individuality and emotion, the rebellion against tradition and rationality, and the infusion of spiritual and supernatural elements. Perfect for a character we all know and love. Enter the moon, stage right.

The Romantics were the rock stars of their day and behaved accordingly, sex, drugs and moonlit boat trips. Central figures such as Percy Bysshe Shelley, Mary Shelley, Lord Byron and John Keats epitomised the Romantic ideals through their exaltation of nature, individualism, and the contemplation of the human condition. Shelley's poetry explored the ephemeral nature of human achievements against the backdrop of the timeless natural world. Lord Byron—a bisexual with a moon-sized sexual appetite, created darkly charismatic heroes and introspective poetry, delving into themes of love, rebellion, and the quest for meaning.

So We'll Go No More A Roving
by Lord Byron (George Gordon)

So, we'll go no more a roving
　So late into the night,
Though the heart be still as loving,
　And the moon be still as bright.

For the sword outwears its sheath,
　And the soul wears out the breast,
And the heart must pause to breathe,
　And love itself have rest.

Though the night was made for loving,
　And the day returns too soon,
Yet we'll go no more a roving
　By the light of the moon.

An actual romance was central to the Romantic movement. It began when the already married Percy Shelley eloped with young Mary Wollstonecraft after threatening suicide if it didn't happen. They had a tempestuous marriage that included holidays with Lord Byron on Swiss lakes. One stormy afternoon, Lord Byron suggested they all write a book to fill in the time between boating, boozing and whatever other nefarious activities they got up to. Byron's look and lifestyle have been compared to a Vampire's. Indeed he started writing a Vampire story, but gave up quickly. However another member of their party, John Polidori, eventually ran with the idea and published The Vampyre (1819). It was a very productive Gothic book club.. Mary (now Mary Shelley) would use this opportunity to write *Frankenstein* (1818), the story of a monster created by a mad scientist from reassembled corpses... a hideously self-aware creature that tended to get around in the moonlight.

> *"One secret which I alone possessed was the hope to which I had dedicated myself; and the moon gazed on my midnight labours, while, with unrelaxed and breathless eagerness, I pursued nature to her hiding-places."*
>
> *Frankenstein, 1818*

In Mary's book, we are often inside the monster's tormented head as it goes on its night journeys, aiming for revenge against its own creation. The moon, a pervasive symbol in Romantic literature, represented the wild, the unattainable, the mysterious, and the connection between the natural world and the inner emotional landscape of individuals.

Bram Stoker, though often associated with the later Gothic tradition through his novel *Dracula* (1897), also drew on Romantic themes of the supernatural and the exploration of human depths. *Dracula* is now the definitive Gothic novel—it's based on the vampire's long history in Eastern European folklore and mythology, refashioned in nineteenth century by the English Romantic gang.

> *"How blessed are some people, whose lives have no fears, no dreads; to whom sleep is a blessing that comes nightly, and brings nothing but sweet dreams."*
>
> *Dracula, 1897*

Dracula changed the vampire's image from a blood sucking bat to a blood sucking gentleman philosopher with sex appeal. It was a hit book in the late nineteenth century and is now one of the world's best known works of English literature. Best read in bed before a restless sleep.

> *Dracula: Mina, to walk with me, you must die to your breathing life and be reborn to mine.*
> *Mina: You are my love... and my life, always.*
> *Dracula: Then, I give you life eternal. Everlasting love.*
>
> Dracula, 1897

Fate

The moon's supernatural influence is also believed to extend beyond the individual to the collective, affecting the fate of nations and the course of natural events. Eclipses, in particular, have been viewed as omens, portending significant changes or cataclysms. This belief in the moon's power to presage events underscores its role as a mediator between the known and the unknown, the natural and the supernatural, serving as a guide and guardian through the mysteries of existence.

The moon, then, is not only a celestial body orbiting the Earth but a symbol of the interconnectedness of all things, a reminder of the unseen forces that permeate the universe. Its supposed supernatural influence offers a lens through which to explore the depths of the human psyche and the mysteries of the cosmos, bridging the gap between the tangible and the ethereal, the scientific and the spiritual. In this way, the moon continues to captivate the human imagination, holding a mirror to our deepest fears and highest aspirations, and inviting us to contemplate the infinite possibilities that lie beyond the boundaries of the known world.

The moon as a symbol of fascination and fear

In conjunction with its enigmatic glow and celestial mystery, the moon has long stood at the crossroads of human emotion, embodying both our deepest fascinations and our primal fears. This duality in perception reflects the complex relationship humanity harbors with the unknown, where the allure of discovery coexists with the trepidation of what lies beyond the familiar. Through the ages, the moon has been a canvas for humanity's imagination, a source of inspiration for myths, legends, and tales that seek to explain the unexplainable, to give form to the shadows that dance in the dark.

The fascination with the moon is as old as time itself, rooted in its omnipresence and its influence over the Earth, from the tides that ebb and flow to the cycles that dictate the natural world. The moon, ever-changing yet constant, has beckoned humanity to look upwards and wonder—to dream of what lies beyond our reach. The moon's phases, its eclipses, and its singular beauty in the night sky have been sources of inspiration for poetry, art, and science, driving the curious and the brave to seek understanding, to navigate the vast oceans, and eventually, to reach its surface. The quest to uncover the moon's secrets is a testament to humanity's insatiable quest for knowledge, a symbol of our desire to transcend our limitations and to explore the unknown.

Behind stories of beasts and apparitions, ancient tales hint at future concepts like the theory of gravity. And so we finish the chapter on a famous moon poem by the British romantic—Percy Shelley—one of the tragic heroes of this nineteenth century movement that questioned the new gods of science and industry. He never achieved fame in his lifetime and drowned in a boat accident in 1822 at the age of 29, when a storm struck. After his body was found, his wife Mary kept his heart wrapped in a silk shroud. The literary fame of the couple grew in the intervening years. In 1852, a year after Mary died, Percy's heart was found in her desk. It was wrapped in the pages of one of his last poems.

Adonais: An Elegy on the Death of John Keats
by Percy Bysshe Shelley, 1821

XII

    Another Splendour on his mouth alit,
    That mouth, whence it was wont to draw the breath
    Which gave it strength to pierce the guarded wit,
    And pass into the panting heart beneath
    With lightning and with music: the damp death
    Quench'd its caress upon his icy lips;
    And, as a dying meteor stains a wreath
    Of moonlight vapour, which the cold night clips,
It flush'd through his pale limbs, and pass'd to its eclipse.

LV

    The breath whose might I have invok'd in song
    Descends on me; my spirit's bark is driven,
    Far from the shore, far from the trembling throng
    Whose sails were never to the tempest given;
    The massy earth and sphered skies are riven!
    I am borne darkly, fearfully, afar;
    Whilst, burning through the inmost veil of Heaven,
    The soul of Adonais, like a star,
Beacons from the abode where the Eternal are.

CHAPTER 7

Howling: a moonlit madness unravelled

The expression 'howling at the moon' brings forth an instant mental image of wolves, crazy people and something ominous. As we will see, right up to the present day, some people take this literally and will gather on a full moon to make a lot of noise. This writer recalls an anecdote from a friend who had started to work in the mental health profession. She said that the hotline would light up on a full moon, and she'd be a lot busier. What is this all about? Does the moon have a real effect on our mental state? Is this just because we are indoctrinated with the moon madness story from early in life? Let's go back in time, when the moon was just like it is today, minus a few American flags, some reflectors, some footprints and some boots. Yes, good readers, there's plenty of other human made stuff there too.

For centuries, the moon has been a source of wonder, inspiring countless myths, legends, and scientific inquiries. Its ethereal presence and cyclical nature have led many to speculate about its potential influence on human behaviour and mental health. In this exploration, we look at the enigmatic relationship

between the moon and madness, uncovering the layers of belief, evidence, and doubt that surround this celestial mystery.

A historical perspective: a short history of lunacy and lore

So now let's look at a quick chronological history of the study of moon madness, looking at the conclusions people have reached over the ages. The term 'lunacy' itself, derived from the Latin word for moon, *luna*, reflects the age-old belief in the moon's power to affect the mind. Ancient civilisations, from the Romans to the Greeks, held the moon in high regard, attributing various psychological changes to its phases. The intricate dance between the moon and madness has been choreographed through the annals of history, intertwining human behaviour with the celestial ballet of our lunar companion.

Antiquity's observations—the moon's mesmeric influence: The ancient Babylonians, as early as 3000 BCE, meticulously recorded the movements of celestial bodies, including the moon, linking its phases with earthly events and human behaviour. These early civilisation's scholars posited that the moon, with its commanding presence in the night sky, wielded significant influence over human affairs, a belief that permeated many aspects of their societal and religious practices.

Greek enlightenment—lunar philosophy and medicine: By the fifth century BCE, the Greeks had begun to embed the moon within the fabric of their philosophical and medical theories. Hippocrates, often hailed as the father of medicine, suggested that "one who is seized with terror, fright, and madness during the night is being visited by the goddess of the moon." This period marked a pivotal moment when lunar effects on mental health began to be considered through a quasi-scientific lens, blending observation with speculation.

Medieval madness—the moon's grip on the mind: Fast forward to the Middle Ages, from the eighth to the fifteenth centuries, the moon's influence over mental health was widely accepted, often used to explain erratic or inexplicable behaviour. The term "lunatic," derived directly from the Latin *lunaticus*, meaning 'of the moon,' became a common label for those suffering from mental illness, reflecting the prevailing belief in celestial causation. Legal documents from England in the thirteenth century even introduced the term "lunacy" to describe insanity believed to be affected by the moon's phases, laying the groundwork for the first lunacy laws which considered the effects of the moon in legal judgements.

Renaissance revelations—celestial bodies and human behaviour: During the Renaissance, a period of rebirth in arts, science, and culture spanning the fourteenth to the seventeenth centuries, luminaries like Paracelsus challenged traditional views, proposing more nuanced relationships between the moon and mental health. Paracelsus, a Swiss physician and alchemist of the early sixteenth century, argued that while the moon might influence the human body and mind, its effects were part of a more complex interplay with natural forces, pushing the boundary between mysticism and emerging scientific thought.

Enlightenment and beyond—scepticism and science: The Enlightenment of the seventeenth and eighteenth centuries heralded a shift towards rationalism and scientific inquiry, leading to increased scepticism about the moon's influence on madness. However, this period did not completely discard the lunar connections; instead, it laid the foundation for future scientific investigations, seeking evidence over anecdotes. It was a time of transition, where the fascination with the moon's potential effects on mental health persisted, albeit under the scrutinising gaze of emerging scientific methodologies.

Scientific inquiry: Separating fact from folklore

As humanity progressed into the age of reason and scientific exploration, the mystical ties between the moon and madness began to be examined through a more empirical lens.

The dawn of empirical scepticism: The late nineteenth and early twentieth centuries marked a pivotal era in which the scientific method began to be applied to the study of human behaviour and mental health. Pioneers in psychiatry and psychology scrutinised the anecdotal evidence linking lunar phases with episodes of madness, seeking empirical data to support or refute such claims. One of the first documented studies, conducted in the late 1800s, aimed to correlate the incidence of psychiatric admissions with the lunar cycle, yielding inconclusive results that spurred further investigation.

Mid-twentieth century studies—seeking patterns in the lunar cycle: By the mid-twentieth century, with advancements in statistical analysis and a growing body of psychiatric knowledge, researchers embarked on more sophisticated studies. These investigations sought to identify patterns in hospital admissions, incidents of violence, and changes in mood or behaviour that could be linked to specific phases of the moon. Despite these efforts, results remained mixed, with some studies suggesting a slight correlation, while others found no significant relationship, highlighting the challenges in isolating lunar effects from other environmental and social factors.

The modern era—advanced methodologies and broader perspectives: The advent of advanced statistical tools and methodologies in the late twentieth and early twenty-first centuries has allowed for more nuanced exploration of the moon's potential effects on mental health. Researchers have expanded their inquiries to include not just psychiatric admissions but also sleep patterns, emergency room visits, and even the timing of births, seeking correlations with the lunar cycle. Meta-analyses of these studies generally suggest that any lunar effect, if present, is likely to be small and not of clinical significance. However, the fascination with lunar influences persists, inviting ongoing scrutiny.

Psychological and biological underpinnings: Beyond mere correlation, scientific inquiry has also ventured into the possible psychological and biological mechanisms through which the moon might exert its influence. Theories have been proposed regarding the moon's impact on the human circadian rhythm, gravitational effects on brain fluid, and even the psychological impact of lunar brightness on human behaviour. While intriguing, these hypotheses remain largely speculative, with conclusive evidence elusive amidst the complex interplay of biological, environmental, and psychological factors.

A continuing quest for understanding: Each study, whether affirming or challenging the lunar connection, contributes to a deeper comprehension of human psychology and its susceptibility to external influences. As we stand on the threshold of new discoveries, the dialogue between ancient lore and modern science continues, reminding us of the enduring mystery that surrounds our celestial neighbour.

It's an effort to sift myth from reality, highlighting the intricate relationship between belief and evidence, underscoring the complexity of human behaviour and the myriad influences that shape our mental health, inviting us to remain open to the wonders of the natural world and the ongoing pursuit of knowledge.

The lunar cycle—phases of influence: The moon's cycle is divided into four primary phases: new moon, first quarter, full moon, and last quarter. Each phase has been associated with specific shifts in mood and behaviour, with the full moon often singled out as a time of heightened emotional sensitivity and erratic behaviour. Like star signs, there is much that doesn't quite make sense about all this... and yet...

New moon—the veil of darkness and renewal: The new moon, a time when the moon is most closely aligned with the Sun and invisible to the Earth, has historically been associated with new beginnings and introspection. In the realm of mental health, this phase is sometimes thought to signify a period of lower emotional energy and a time for reflection. Scientific investigations into this period have sought to understand whether there is a decrease in psychological disturbances or a subtle shift in mood patterns, yet findings have remained largely inconclusive, with no significant spikes in hospital admissions or psychiatric incidents reliably linked to this lunar phase.

First quarter—the waxing crescent of building tension: As the moon waxes towards the first quarter, symbolising growth and accumulation, there has been speculation about its corresponding increase in emotional tension and stress among individuals. Researchers have explored the possibility of a gradual build-up in psychiatric cases or disturbances during this phase, hypothesising that the increasing lunar visibility might affect human behaviour. While some anecdotal evidence suggests a slight uptick in anxiety or emotional unrest, comprehensive studies have yet to establish a definitive correlation, leaving the door open for further exploration.

Full moon—the apex of light and lore: The full moon phase, when the moon is fully illuminated by the sun, stands at the heart of lunar folklore, often cited as a time of heightened emotional intensity, increased psychiatric admissions, and amplified erratic behaviour. This period has been the focus of numerous scientific studies aiming to uncover the truth behind the claims of "lunar madness." While the romantic notion of the full moon's power over human behaviour persists in popular culture, systematic reviews and large-scale analyses have predominantly found no substantial evidence to support a significant increase in mental health crises or hospital admissions correlated with the full moon.

And yet... again, this writer has heard personal anecdotes that contradict these studies. It is also common to hear people naming the full moon as the reason for some behaviour or incident. This quote is from a 2024 interview with Manda Kaye, an Australian moon enthusiast.

> *Manda Kaye: "I used to be a high school teacher and, I swear to God, we used to always say 'oh my God, it's a full moon and the kids are feral'."*

Last quarter—the waning light and release: Following the peak illumination of the full moon, the waning phase towards the last quarter represents a period of release, where it is often believed that emotional and psychological tensions begin to ebb. The diminishing lunar light is sometimes associated with a decrease in the intensity of psychological disturbances, symbolising a return to equilibrium. Scientific scrutiny of this phase has similarly struggled to find clear patterns of behavioural change or psychiatric incidents that could be attributed directly to the lunar cycle, further complicating the narrative of the moon's influence on mental health.

Bridging belief and evidence

Across each phase of the lunar cycle, the quest to delineate the moon's influence on human psychology intertwines centuries-old beliefs with contemporary scientific inquiry. While the allure of the moon's phases continues to inspire fascination and speculation, the empirical evidence remains elusive, painting a complex picture of the interplay between celestial phenomena and human behaviour. This ongoing dialogue between myth and methodology not only enriches our understanding of the lunar cycle's cultural significance but also challenges us to critically evaluate the forces that shape our perceptions of mental health and wellness.

The exploration of the lunar cycle and its phases of influence reveals a landscape where folklore and fact converge, offering a multifaceted view of humanity's relationship with the moon. As science advances, the pursuit of knowledge about the moon's impact on us—psychologically, emotionally, and behaviourally—continues to evolve, reflecting our perpetual fascination with the cosmos and its mysteries. Even if people's behaviour is based on myth and rumour—many people will behave differently according to the moon.

> *Don't blame it on sunshine*
> *Don't blame it on moonlight*
> *Don't blame it on good times*
> *Blame it on the boogie.*
>
> *Michael Jackson, 1978*

Psychological perspectives: Moonlit reflections

The quest to understand the moon's influence on human psychology transcends the search for empirical evidence, venturing into the realm of psychological interpretation and theoretical exploration.

The moon and the mind—an intriguing connection: The fascination with the moon's influence on human psychology is not merely a matter of correlating lunar phases with behavioural patterns; it involves understanding how the moon's presence might resonate with the human mind on a deeper, more symbolic level. Psychologists have explored the moon as a metaphor for the unconscious, a luminous body illuminating the night and potentially bringing hidden desires and fears to the surface. This symbolic relationship suggests that the moon's phases might reflect internal psychological cycles, offering a mirror to our emotional ebb and flow.

Circadian rhythms and sleep—the impact of the lunar cycle: One of the most concrete areas of interest is the moon's influence on human circadian rhythms and sleep patterns. Studies have investigated whether the full moon's increased luminosity disrupts sleep, potentially leading to changes in behaviour or mood.

While research findings have been mixed, some studies suggest that people may experience slight alterations in sleep quality around the full moon, which could, in turn, affect emotional and psychological wellbeing. This line of inquiry highlights the complex relationship between environmental cues, biological rhythms, and psychological states.

Social behaviour and collective psychologies: Beyond individual psychological effects, some theories propose that the moon influences collective behaviours and societal mood swings. During certain lunar phases, particularly the full moon, there is a belief that social interactions may intensify, leading to increased communal tension or euphoria. Although hard evidence supporting this theory is scarce, the notion persists that the moon can sway the collective psyche, amplifying social dynamics and emotional responses within communities. The 1980s saw full moon raves gain popularity all over the world. While drugs played a big part in them, the moon was an integral element.

The psychological mechanisms behind lunar beliefs: Psychologists also examine why beliefs in the moon's influence remain so pervasive, despite the lack of consistent scientific support. Cognitive biases, such as confirmation bias, may play a role, where individuals are more likely to notice and remember events that confirm their beliefs about the lunar effects. Additionally, the moon's cultural and historical significance contributes to its psychological impact, weaving lunar lore into the fabric of societal narratives and individual belief systems.

Embracing the mystery—the moon as psychological muse: Ultimately, the moon serves as a powerful psychological muse, inspiring wonder, curiosity, and introspection. Whether or not its phases have a direct impact on mental health, the moon's symbolic presence in human culture encourages a deeper exploration of our emotional and psychological landscapes. It invites us to consider the broader influences on our behaviour and wellbeing, from the environmental to the existential, and to remain open to the mysteries of the natural world and the human mind.

In late nineteenth century Germany, the human unconscious became an area where a new kind of doctor, called a psychiatrist, began to dig in order to heal. Carl Jung was a disciple of Sigmund Freud, and took Freud's new medical science of psychoanalysis into the lunar zone. Jung was a big fan of symbolism, the collective unconscious and looking at the deeper meanings of dreams.

There is a desert on the moon where the dreamer sinks so deeply into the ground that she reaches hell.

Carl Gustav Jung, 1961

Scepticism and critique: A rational look at lunar effects

But not everyone was convinced by Jungian psychology or the whole moon madness correlation. In the quest to demarcate the boundary between lunar myth and mental health reality, a wave of scepticism emerged, advocating for a rational, evidence-based approach to understanding the moon's influence on human psychology.

The challenge of empirical evidence: Central to the sceptics' arsenal is the argument concerning the lack of robust empirical evidence linking the lunar cycle to significant changes in human behaviour or mental health outcomes. Despite numerous studies conducted over the decades, a consensus remains elusive, with many researchers concluding that any correlation between the moon's phases and psychological disturbances is minimal or non-existent.

Critics argue that the persistent belief in lunar effects is more a testament to the power of folklore and anecdote than to any scientifically validated phenomenon.

Debunking lunar myths—the role of confirmation bias: One of the primary psychological mechanisms sceptics cite in explaining the enduring belief in lunar effects is confirmation bias. This cognitive bias leads individuals to notice, remember, and interpret information in a way that confirms their preconceptions, effectively overlooking evidence that contradicts their beliefs. Sceptics contend that anecdotal reports of increased erratic behaviour or psychological distress during the full moon are often highlighted, while uneventful lunar phases go unreported, creating a skewed perception of the moon's impact.

Statistical analysis and the full moon fallacy: Critics of lunar influence theories also point to the misinterpretation of statistical data, often referred to as the "full moon fallacy." They argue that when large datasets are analysed—spanning emergency room visits, psychiatric admissions, and police reports—no significant deviation emerges during the full moon compared to other lunar phases. This statistical scrutiny challenges the anecdotal evidence and highlights the importance of rigorous data analysis in debunking myths surrounding lunar effects on mental health.

The psychological and social underpinnings of lunar lore: Beyond the critique of empirical evidence, sceptics also explore the psychological and social reasons behind the persistence of lunar lore. They suggest that the human propensity for pattern recognition and narrative construction plays a crucial role in perpetuating the myth of the moon's influence.

The moon's visible changes and its prominent place in cultural and historical narratives make it a compelling figure for attributing causality to unexplained or random events, reinforcing its mythical status in the collective consciousness.

Moving beyond the moon: A call for scientific rigour

The sceptical perspective calls for a move beyond mystical attributions to the moon, urging a more scientifically rigorous approach to understanding human psychology and behaviour. By advocating for evidence-based conclusions, sceptics aim to shift the focus from celestial phenomena to more grounded psychological and environmental factors that genuinely influence mental health. This rationalist view does not diminish the cultural or symbolic significance of the moon but encourages a more nuanced appreciation of the complexities of human behaviour, free from the shackles of lunar mythology. These sceptics can sometimes sound like a collective group of doubters who exist in a permanent state of a waxing moon, but let's move on to some personal stories.

Personal stories: Moonlit testimonies

Amidst the scientific debates and skeptical critiques, the moon's influence on human behaviour and mental health continues to be a subject of personal testimony and anecdotal evidence. Literature is full of obviously personal anecdotes about the moon.

> ***Theseus:***
> *Now, fair Hippolyta, our nuptial hour*
> *Draws on apace. Four happy days bring in*
> *Another moon—but, Oh, methinks how slow*
> *This old moon wanes! She lingers my desires,*
> *Like to a stepdame or a dowager*
> *Long withering out a young man's revenue.*
>
> ***Hippolyta:***
> *Four days will quickly steep themselves in night;*
> *Four nights will quickly dream away the time;*
> *And then the moon, like to a silver bow*
> *New bent in heaven, shall behold the night*
> *Of our solemnities.*
>   William Shakespeare, 'A Midsummer Night's Dream', 1595

And as we observed in the previous chapter, the English Romantic movement was full of moon symbolism.

> *I*
> *Art thou pale for weariness*
> *Of climbing heaven and gazing on the earth,*
> *Wandering companionless*
> *Among the stars that have a different birth,—*
> *And ever changing, like a joyless eye*
> *That finds no object worth its constancy?*
>
> *II*
> *Thou chosen sister of the Spirit,*
> *That gazes on thee till in thee it pities ...*
> *Percy Shelley, 'To the Moon', 1820*

These writers are using the moon as a metaphor because it is such a universal experience. The moon means something to everyone on Earth. They may have different experiences but there are common elements.

The full moon's emotional resonance

Numerous individuals report a distinct change in their emotional landscape during the full moon. Stories often describe heightened feelings of restlessness, increased sensitivity, and a greater propensity for emotional upheaval. For some, the full moon is a time of introspection and heightened creativity, where the mind seems more open to exploring the depths of thoughts and feelings. These testimonies, while subjective, underscore the profound connection many feel with the lunar cycle, suggesting a synchronicity between the moon's phases and personal emotional cycles.

Sleep patterns under the lunar glow

Another common theme in personal accounts is the impact of the moon, particularly the full moon, on sleep quality. Individuals recount tales of restless nights, difficulty in falling asleep, or unusually vivid dreams during this phase. While scientific studies offer mixed findings on the moon's influence on sleep, these personal experiences highlight the perceived direct impact of lunar light and its phases on human circadian rhythms, contributing to the ongoing dialogue about the moon's role in our lives.

Lunar tides in mental health

Beyond sleep and emotional states, some narratives draw a connection between the lunar cycle and more significant mental health challenges. People with mood disorders, for example, have shared observations of a fluctuation of their symptoms, which have aligned with the phases of the moon. Though such correlations lack consistent scientific backing, they provide a compelling glimpse into the ways individuals seek to understand and contextualise their experiences within the broader rhythms of the natural world.

1. The effects of the full moon on mental wellbeing and madness were usually characterised in popular media through images of the asylum.
2. The Blackheath Howlers in Sydney's Blue Mountains region.

Collective behaviours and the lunar influence

Tales of the moon's influence extend beyond the individual to encompass community and societal dynamics. During certain lunar phases, especially the full moon, communities may report an uptick in social gatherings, celebrations, or communal tensions. These collective experiences, passed down through generations or shared among peers, reinforce the cultural lore surrounding the moon, embedding its presence in the social fabric of human interaction.

Modern moon howlers

And here, we detour to a scenic suburb in the Blue Mountains, near Sydney, Australia in the year 2024. The Blackheath Howlers are a loose group that gather on full moons and howl. Why would you do this?

> *Manda Kaye, Blackheath Howlers:* "*I moved up here in 2011 and it became a bit of a thing that we did. People came up from Sydney, and we would take them to Govetts Leap, which is an absolutely spectacular lookout. We'd go there just for drinks at first. We called it 'howlers' because we were getting together for the full moon and we thought—well, what happens at the full moon? Werewolves come out, dogs howl. So it just felt natural to call it howlers. I can't remember exactly when we did it, but we started actually howling and people really liked it. You can come down and take all of the woes of the last month and put them out there into the valley. Howl into the abyss—as it were.'*

The Blackheath Howlers have a Facebook group and regularly get out there to howl at the moon. For Manda, it's not particularly 'woo woo', it's just a bit of fun.

> "It's very therapeutic. People love it. A couple of months ago, somebody brought their 90-something year old mum. She wasn't sure there was going to be howling and didn't quite believe it. But she was well into it, she was having a great time!"

For Manda, the full moon makes you think back to all the people who've witnessed the same show over the eons.

> "It's never boring. Recently, it fell on Invasion Day (26 January, Australia Day) and we were very conscious that we're doing this on Darug and Gundungurra land. We do think about how many people stood there and looked at the moon before and marveled at the same place. It's also somewhere Charles Darwin visited. And scholars suggest it informed his theory of evolution. Because he looked at the stratified valley sides and went—you know—this is not 2,000 years old. So it's a site that I associate with the spiritual and also science."

Like the chapters in this book, there are different ways of looking at the same magnificent orb. In the quest to understand the moon's effects on madness, we find ourselves navigating a labyrinth of myth, science, and personal experience. While definitive answers may remain elusive, it's fun to stare up and think about them. The moon's effect on madness could be real or imagined, but if so many people imagine it—then it is real!

CHAPTER 8

The relationship between the moon and the tides

Our wonderful moon, not only a poet's inspiration but a cosmic DJ spinning Earth's very own oceanic turntable. Yes, you heard that right! When it comes to Earth's club mix, the moon drops the beat with its gravitational pull, orchestrating the rise and fall of sea levels in what we know as tides. This isn't just your average celestial phenomenon; it's a tidal force that has the power to move oceans, literally.

Imagine the moon with its invisible gravitational strings, plucking at Earth's blue surface, causing the waters to dance in a rhythmic rise and fall. This celestial interaction is most lively during the full and new moon phases, where the tides hit their peak performance, known as spring tides. And when the moon decides to pump up the volume, and the oceans respond with higher highs and lower lows.

But the moon's tidal remix doesn't just make for an epic natural spectacle; it plays a special role in the health and sustainability of marine ecosystems. From the breeding grounds of the deep sea to the bustling nurseries of coastal wetlands, the ebb and

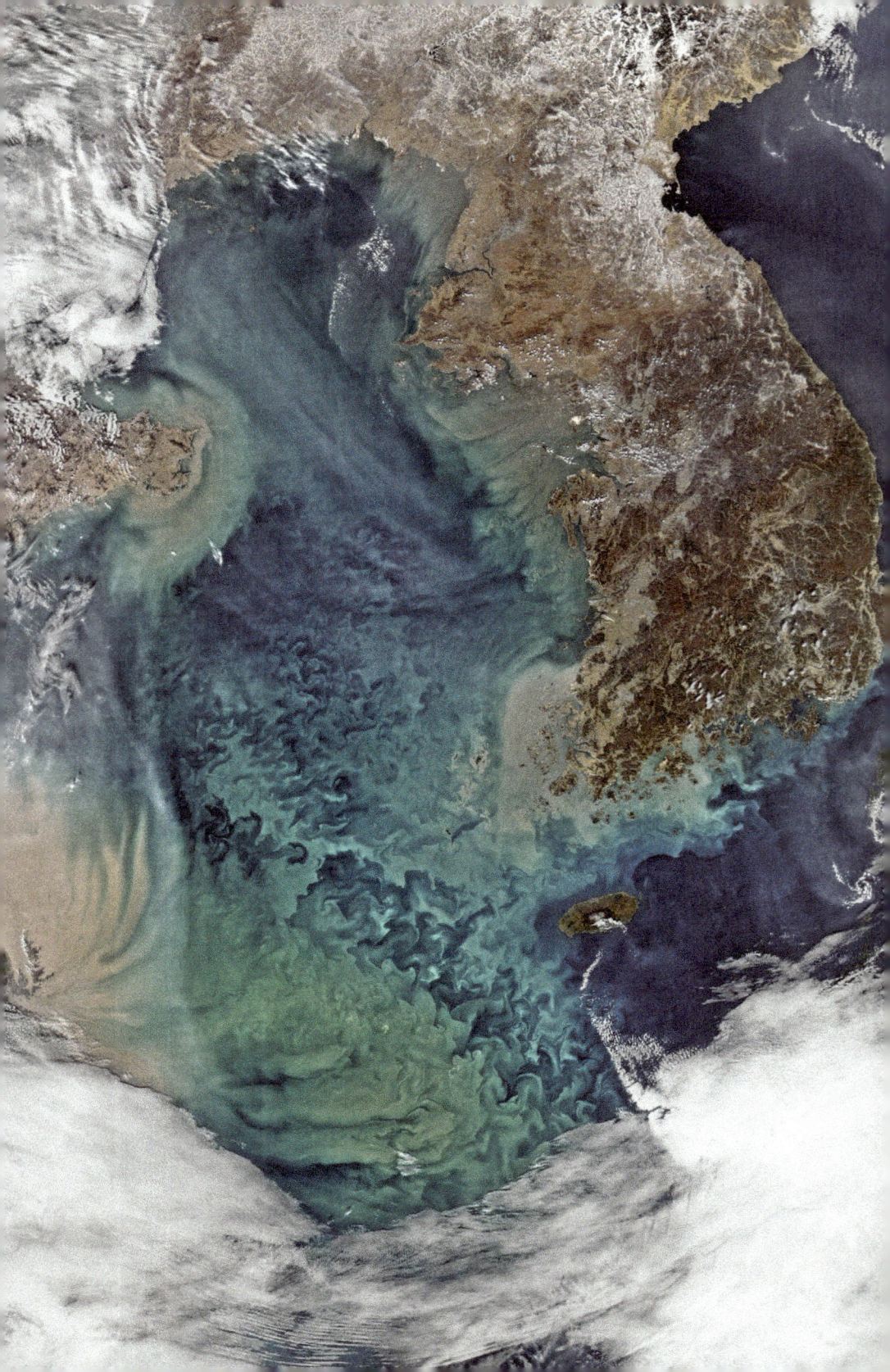

flow of tides, under the moon's guidance, influence the life cycles of countless marine organisms. It's as if the moon is conducting the orchestra of marine life, ensuring each section comes in at the right time for the health of the ecosystem.

Not to be overlooked is the moon's influence on human activities. Since ancient times, navigators have used the moon's phases to predict tidal flows, charting safer and more efficient paths across the seas. It's the original GPS, guiding ships long before satellites took over. And let's not forget the coastal cultures, for whom the moon and tides are not just elements of nature but pivotal characters in myths, legends, and daily life. From the fishing calendars set by the tidal schedule to the festivals celebrating the lunar influence, the pull of the moon is there.

And while Earth's tides are a signature tune, the moon's gravitational gig isn't a one-planet show. Across the solar system, other moons perform their own gravitational symphonies with their host planets, each with unique effects—from volcanic eruptions on Jupiter's moon Io caused by tidal heating to the potential future ring system of Mars, courtesy of Phobos's inward spiral. These interplanetary performances offer a broader understanding of how gravitational forces play out on a cosmic scale, reminding us that Earth's own tidal drama is part of a much larger celestial concert.

The moon is Earth's partner in a cosmic spectacle that influences the very rhythm of life on our planet. From guiding the paths of ancient mariners to shaping the biological rhythms of marine species, the Moon's gravitational pull is a force that weaves together the fabric of Earth's natural and human spheres. It's a reminder of the interconnectedness of all things celestial and terrestrial, and the profound impact of our nearest cosmic neighbour on life as we know it. Our silvery moon may have even been partly responsible for the birth of organic life on this planet. And taking this thought even further—for shaping our intelligence.

The weird force of gravity

Gravity is a strange, fundamental force of nature that attracts two bodies toward each other. It's the force that causes objects to fall to the ground when dropped, keeps the planets in orbit around the Sun, and governs the motion of galaxies and clusters in the universe. The strength of the gravitational force between two objects depends on their masses and the distance between them: it increases with mass and decreases with distance.

The English mathematician, physicist, and astronomer Isaac Newton, formulated the theory of gravity in the late seventeenth century. His groundbreaking work, *Philosophiæ Naturalis Principia Mathematica* (Mathematical Principles of Natural Philosophy), first published in 1687, laid the foundation for classical mechanics. Legend has it that the theory came to him when an apple fell on his head, although this has been largely discredited. But perhaps he was watching that big yellow ball outside the study window on a cold night, wondering why it didn't fall, then thinking about the even bigger ball he was sitting on? Whatever did happen, we'll never know but Newton came up with the universal law of gravitation, stating that every point mass attracts every other point mass by a force acting along the line intersecting both points. This force is proportional to the product of their masses and inversely proportional to the square of the distance between them. Newton's law of universal gravitation was revolutionary because it applied both to celestial bodies and to objects on Earth, providing a unified description of gravity.

THE RELATIONSHIP BETWEEN THE MOON AND THE TIDES

Newton's theory of gravity explained why planets orbit the Sun in elliptical paths, a phenomenon that had been observed by astronomers such as Johannes Kepler. Newton's equations also explained the tides by showing how the gravity of the moon influences the Earth's oceans, a topic of significant interest and debate before his work.

The moon, Earth's only natural satellite, played a crucial role in demonstrating Newton's theory of gravity. The gravitational pull between the Earth and the moon not only stabilises the moon's orbit around Earth but also causes the tides. The moon's gravity pulls on the Earth's oceans, causing high and low tides in a continuous cycle. The successful landing of *Apollo 11* on the moon in July 1969 offered a unique opportunity to study gravity in a different environment. The astronauts' ability to jump higher and carry heavier loads than on Earth was a direct consequence of the moon's weaker gravitational pull, which is about 1/6th that of Earth's, and this difference is due to the moon's smaller mass compared to Earth.

Newton's universal law of gravitation has stood the test of time, accurately describing the gravitational interactions for centuries. However, the theory was later expanded by Albert Einstein's theory of general relativity in the early twentieth century, which provided a more comprehensive description of gravity, especially under conditions of extreme mass and gravity, such as near black holes where Newton's laws do not suffice. Despite this, Newton's theory remains a fundamental principle for understanding and calculating gravitational forces in most contexts, including space exploration and satellite technology.

Then there is the unnerving way that the moon keeps one face towards the Earth. It's like the spin force has been frozen by the Earth's gravity! While the Earth rotates away beneath it… hence, the expression 'dark side of the moon', the side we never see from Earth. There are mind-blowing reasons for this—the moon's rate of spin is tidally locked so that it is synchronised with its rate of revolution. It's a different way of thinking of the tidal force to the watery version, but the same thing. The result is the moon rotates exactly once every time it circles the Earth.

It gets weirder: gravitons

What makes gravity act at a distance? Gravitons are hypothetical elementary particles that mediate the force of gravity in the framework of quantum field theory. Unlike the classical view of gravity described by Newton's law of universal gravitation and further refined by Einstein's theory of general relativity, the concept of gravitons comes from attempts to formulate a quantum theory of gravity. In the quantum field theoretical approach, forces are mediated by particles: for electromagnetism, the force carrier is the photon; for the weak nuclear force, it's W and Z bosons; and for the strong nuclear force, it's gluons. Similarly, the graviton is proposed as the quantum of the gravitational field, acting as the force carrier for gravity.

But do they actually exist? If they did, they would be massless (because gravity has an infinite range) and would travel at the speed of light. They would also be spin-2 particles, which means they would have a quantum spin that helps determine how they interact with matter. The idea is that when mass distorts spacetime, as described by general relativity, gravitons would be the particles that mediate the resulting gravitational force between masses. Essentially, the exchange of gravitons between two masses would be what causes what we perceive as the gravitational attraction between them.

However, it's important to note that gravitons have not been detected or observed directly. The concept remains theoretical, and detecting a graviton is a formidable challenge because the gravitational force is incredibly weak compared to the other fundamental forces. This weakness means that the effects of individual gravitons are minuscule, making experiments to detect them directly or indirectly extremely difficult with current technology.

The search for gravitons and a successful theory of quantum gravity remains one of the major unsolved problems in physics. A theory that successfully integrates quantum mechanics with gravity would not only have to explain how gravitons work but also reconcile the principles of quantum mechanics with the curved spacetime description of gravity provided by general relativity. Such a theory, often referred to as a theory of quantum gravity, would represent a significant breakthrough in our understanding of the universe, potentially unlocking new insights into the nature of space, time, and the fabric of reality itself. It's possible that humans are simply incapable of grasping the concepts required to understand these things. But Newton, Einstein and the group of minds who formulated quantum mechanics are proof that it is possible to think through abstract concepts and make them comprehensible. At least for a select group of earthbound thinkers. Even today, there is disagreement in the physics community about how many people truly understand Einstein's General Theory of Relativity 'intuitively'… are you one of them?

The gravitational influence of the moon

Meanwhile, out in near space, the moon continues to provide solid evidence of gravitational force at a distance. The force exerted by the moon on Earth is not uniform; it varies across the planet's surface. This variation is due to the difference in distance between the moon and different parts of the Earth. The side of Earth closest to the moon experiences a stronger gravitational pull, leading to a bulging of water towards the moon, which we observe as high tide. Conversely, on the side of Earth farthest from the moon, the gravitational pull is weaker, but inertia causes water to bulge outward, creating another high tide. Between these two high tides, lower sea levels, or low tides, occur. This interplay of gravitational forces and Earth's rotation results in the regular ebb and flow of tides.

The tidal forces generated by the moon have big implications for life on Earth. They influence the marine ecosystems with the tidal movements contributing to the mixing of water, nutrients, and temperature regulation. This mixing is essential for the health of many marine species, affecting breeding, feeding, and migration patterns. For coastal communities and ecosystems, tides shape landscapes, affecting erosion, sedimentation, and the formation of natural habitats like estuaries and mangroves, which are critical for biodiversity.

The moon's gravitational influence on tides has also played a pivotal role in human history and development. For ancient and modern navigators, understanding the patterns of tides has been essential for safe and efficient sea travel. Many cultures have developed calendars and timekeeping systems based on the lunar cycle and its effect on tides. The predictability of tidal movements, governed largely by the moon's phases, has allowed people to plan agricultural, fishing, and hunting activities.

Additionally, in some cultures, the moon and its control over the tides have been imbued with spiritual and religious significance, often celebrated in myths, festivals, and rituals. For Māori, Tangaroa (*or* Takaroa in the South Island of New

THE RELATIONSHIP BETWEEN THE MOON AND THE TIDES

Zealand/Aotearoa) is the great atua (god) of the sea, lakes, rivers, and creatures that live within them, especially fish. Tangaroa is considered to wield immense power over the ocean's domain, influencing not only the tides but also the wellbeing of marine creatures and, by extension, the prosperity of the Māori people themselves. The movements of the tides are interpreted as expressions of Tangaroa's mood and intentions, guiding fishing, navigation, and the timing of rituals. The cyclic nature of the tides also symbolises the interconnectedness of all elements of the natural world, reflecting the Māori belief in the balance and harmony between humans and the environment. This holistic perspective underscores the respect Māori have for the natural world, with the tides serving as a constant reminder of the dynamic and sacred relationship between the people and the sea.

Spring tides

There's another force at work with the tides, the moon's old rival: the Sun. While that overly bright show off is the winner in size and luminosity, the moon is the winner in tidal powers. Yes, the moon exerts roughly twice the force that the sun does over the tides. Why? It's smaller but closer. Here's a quick quiz: the mass of the sun is 1.999×10 to the power of 30 kilograms, the moon is 7.35×10 to the power of 22 kilograms. The moon is on average 383,500 kilometres away from the Earth. How far is the Sun? If you go back to Newton's theory earlier in this chapter, you can work it out for yourself... easy.

Spring tides are the Sun and the Moon working together. As they say: keep your enemies close! The moon's gravitational pull, reaches its zenith during specific lunar phases, namely the full moon and the new moon. During these periods, when there is an alignment of the Earth, moon and Sun, there is a combined gravitational pull on the Earth's oceans, resulting in the phenomena known as 'spring tides'. These tides are characterised by their exceptionally high and low waters, marking a stark contrast to the more moderate tides experienced during other phases of the lunar cycle.

The dynamics of spring tides offer a fascinating insight into the interplay of celestial bodies. When the Sun, Earth, and moon align during the full and new moon phases—events known as syzygy—the gravitational forces of the Sun and moon combine to exert a more substantial pull on Earth's waters. The synergistic effect of these forces enhances the tidal bulges, leading to higher high tides and lower low tides. This phenomenon underscores the delicate balance within our solar system, where distances and alignments of celestial bodies culminate in tangible effects on our planet's natural systems.

The significance of the lunar tide-generating force extends beyond its role in creating spring tides. These heightened tides play a crucial role in the ecological and geological processes of Earth's coastal and marine environments. During spring tides, the increased movement of water can lead to the redistribution of sediments, shaping the landscape of coastal areas and influencing the habitats of countless marine species. For ecosystems reliant on the intertidal zone, the extremes of spring tides provide unique conditions that support a diverse range of life forms, many of which have adapted to thrive in the fluctuating environments between land and sea.

The predictability of spring tides, governed by the lunar cycle, has historically been integral to human societies. Coastal communities, maritime industries, and navigational endeavours have developed with a keen awareness of the timing and impact of these tides. The ability to anticipate the occurrence of spring tides has enabled the planning of activities ranging from fishing and shellfish harvesting to the scheduling of shipping and exploration ventures. This knowledge, passed down through generations, has been instrumental in the survival and prosperity of cultures intimately connected with the sea. As we explore deeper into the mysteries of the universe, the phenomena of spring tides remain a compelling example of the interconnectedness of cosmic forces and terrestrial life.

The impact of tides on marine ecosystems

The moon, through its gravitational pull, orchestrates the tides, crafting a dynamic environment that is crucial for the vitality of marine ecosystems. This regular ebb and flow of tides, a direct consequence of the lunar influence, is not just a phenomenon of rising and falling sea levels but a fundamental ecological force that shapes the lives of countless marine organisms. The rhythmic movement of tides brings about a sequence of environmental changes that are instrumental in influencing the breeding, feeding, and migratory patterns of a diverse array of marine life, underscoring the interconnectedness of celestial dynamics and earthly life.

Marine ecosystems, characterised by their rich biodiversity, rely heavily on the tidal movements for the cycling of nutrients and energy. As tides rise and fall, they churn the waters, facilitating the distribution of nutrients from the depths to the surface, where they become accessible to a wide range of species. This nutrient redistribution is essential for the productivity of plankton, which forms the base of the marine food web. In turn, the abundance of plankton supports larger marine animals, from fish to marine mammals, ensuring a continuous supply of food and maintaining the balance of marine ecosystems.

The influence of the moon on tides also extends to the deeper waters, where tidal currents contribute to the mixing of water layers, affecting temperature and salinity gradients. This mixing is crucial for deep-sea ecosystems, where light does not penetrate, and life relies on the slow but steady rain of organic matter from the surface. The currents generated by tides help distribute this organic matter, supporting deep-sea communities and contributing to the overall health of the marine environment.

The predictability of tidal patterns has been a boon for many marine species, allowing them to synchronise their reproductive activities with optimal environmental conditions. For instance, certain species of corals release their eggs and sperm in a synchronised spawning event, timed to follow the full moon. This timing ensures that the maximum number of offspring survives, taking advantage of the currents to spread and settle in new areas.

The impact of the moon on marine ecosystems is a testament to the intricate links between celestial movements and Earth's biological processes. The tides, driven by the moon's gravitational pull, are not merely physical phenomena but vital ecological drivers that shape the distribution, abundance, and diversity of life in the marine realm.

The influence on coastal regions

Erosion and sedimentation are natural processes influenced significantly by the movement of tides. As tides ebb and flow, they carry with them the power to erode coastal landforms, gradually wearing away rocks and cliffs, and to deposit sediments in other areas, creating new landforms over time. This constant reshaping of the coastline is partly driven by the energy of tidal movements in conjunction with the weather. In areas with strong tidal currents, the erosion can be dramatic, leading to the formation of unique geological structures such as sea arches, stacks, and caves. Conversely, in regions where the tide brings in more sediment than it carries away, expansive beaches, sandbars, and deltas can form, altering the coastline and creating new land.

The development of habitats such as estuaries and wetlands is another crucial aspect of the moon's influence on coastal regions. Estuaries, where freshwater from rivers meets and mixes with saltwater from the sea, are particularly dynamic environments that owe their existence and productivity to the ebb and flow of tides. Tidal movements in these areas facilitate the exchange of water, sediments, and nutrients, creating conditions that support a diverse array of plant and animal species.

Wetlands, including salt marshes and mangrove forests, also benefit from tidal action, which provides essential nutrients and sediments that support their complex ecosystems. These habitats are not only biodiversity hotspots but also provide critical services to humans, such as flood protection, water filtration, and carbon sequestration, underscoring the importance of tidal influences in supporting ecological and human wellbeing.

The influence of the moon on coastal regions extends to the human dimension as well. Coastal communities have historically developed in harmony with the rhythm of the tides, relying on them for navigation, fishing, and the cultivation of tidal flats. The management of coastal resources, the design of coastal defences, and the planning of development projects in these areas are all informed by an understanding of tidal patterns and their impact on the coastal landscape.

How the tides affect navigation and influenced human activity

The moon's gravitational pull, while invisible to the naked eye, has been a silent guide for navigators and seafarers throughout human history. Reading the ocean and its movements was key to the incredible ocean voyages undertaken across the Pacific Ocean by Polynesians before any European ships appeared on the horizon. The moon was a useful light and directional tool as well as an influence on the sea itself.

The rhythmic ebb and flow of the tides, governed by the lunar cycle, provide a natural timetable for the departure and arrival of ships. In ancient times, understanding the tidal patterns was essential for crossing shallow waterways and avoiding the dangers posed by rocks and sandbanks that could become exposed or submerged with the tide's change. Mariners learned to time their voyages with the tides, using the high tides to gain access to ports and navigate through treacherous passages safely. This knowledge was passed down through generations, becoming an integral part of maritime lore and tradition.

The predictability of tides extended beyond the immediate concerns of navigation to influence the very infrastructure of maritime communities. Harbours, docks, and canals were often designed with tidal patterns in mind, maximising the natural flow of water to facilitate ship movements and the loading and unloading of goods.

The relationship between the moon, tides, and navigation also played a crucial role in the Western age of exploration. Explorers charting unknown waters relied on their knowledge of tides to venture into unexplored territories, discover new lands, and establish trade routes across the oceans. The moon's phases provided a celestial reference that helped navigators determine their position at sea and make calculated decisions about their voyages. This celestial navigation, combined with an understanding of tidal forces, enabled the great maritime expeditions that connected distant parts of the world, fostering cultural exchange and the spread of knowledge.

In modern times, the significance of tides in navigation has not diminished. Though technological advancements have revolutionised sea travel, the importance of understanding tidal patterns remains. Today's mariners use sophisticated tools and systems that incorporate tidal data for route planning, ensuring safety and efficiency in marine operations. The predictability of tides continues to be crucial for activities such as coastal engineering, where the timing of construction projects must align with tidal cycles to mitigate impacts on marine environments and human communities. Harbour pilots with intimate knowledge of particular locations are regularly sent out to large cruise and cargo ships to bring them in and out of ports.

How understanding the moon and tides has changed humanity

The moon has been key to astronomy and understanding the mechanics of the universe. It's an unavoidable object in the sky—no telescope needed. Since ancient times, the observation of the moon and its phases has been central to our quest to comprehend the cosmos.

Historically, the moon's phases and its impact on the tides were among the first observable phenomena that suggested a profound connection between celestial bodies and the natural world. Ancient civilisations, observing the regular ebb and flow of the tides, began to notice patterns that corresponded with the lunar cycle. These observations were crucial in leading to the understanding that the moon exerts a gravitational pull on the Earth, influencing the movement of the seas. This realisation marked a significant leap forward in human knowledge, bridging the gap between the heavens and Earthly phenomena and laying the groundwork for the field of celestial mechanics.

The moon's influence on tides became a subject of fascination and study for many early astronomers and natural philosophers. In their efforts to understand the cosmos, they saw in the moon's effects on the tides evidence of the invisible forces that govern the movements of celestial bodies. The predictable nature of the tides, aligned with the moon's phases, offered a tangible example of cosmic order, a concept that was central to the development of astronomy as a science. This understanding contributed to the formulation of theories about the gravitational pull of celestial bodies, foreshadowing the groundbreaking work of figures such as Kepler and Newton.

Kepler's laws of planetary motion and Newton's law of universal gravitation were milestones in the history of science, providing a mathematical framework to describe the orbits of planets and the forces that govern them. The moon's role in influencing the tides served as a crucial empirical observation that supported these theories, demonstrating the universal nature of gravitational attraction.

Tides and creatures
The reproductive activities of numerous intertidal species are closely tied to the lunar calendar. The grunion fish of the California coast, for example, come ashore to lay their eggs during the high tides of the full and new moons. The eggs are buried in the sand, where they remain until the next series of high tides, two weeks later, which helps ensure their hatching.

This remarkable strategy demonstrates the profound influence of lunar-driven tides on the reproductive success of marine species.

Migration patterns of marine life are also significantly impacted by the moon and the tides. Many species of fish and marine mammals navigate and time their migrations based on tidal flows, which can be more or less pronounced depending on the lunar phase. The strong currents of spring tides, occurring during the full and new moons, provide cues for movements and migrations. For example, salmon are known to use the high tides as a signal to begin their upstream migration, ensuring that they have sufficient water depth to navigate obstacles.

The foraging behaviours of many marine and coastal animals are intricately linked to the tides, with the lunar cycle dictating the availability of food resources. Shorebirds time their feeding to the low tides when mudflats are exposed, revealing a bounty of invertebrates to feed on. The tidal cycles, governed by the moon, thus become a critical factor in the survival strategies of these species, influencing when and where they feed.

The moon's gravitational pull and its resulting tides also play a crucial role in the distribution of nutrients within marine ecosystems. Tidal mixing helps circulate nutrients from the depths to the surface waters, supporting plankton blooms that form the base of the oceanic food web. This nutrient cycling is essential for maintaining the productivity and health of marine ecosystems, upon which countless species rely.

Money making tides

The economic relevance of tides extends across various sectors, there's fishing and tourism and you can even generate electricity via the moon's tidal powers! The fishing industry, a vital source of food and livelihood for millions of people

1. The waters and the moon. 2. Sir Isaac Newtown musing about the effects of the moon on tides. 3. The best waves can sometimes be found during a full moon. 4. Atlantic Ocean, as seen from the moon. 5. The effects of tides on coral in the Bahamas.

worldwide, relies heavily on the knowledge of tidal patterns. Many marine species, targeted by commercial and subsistence fishers, exhibit behaviours tied to the tides, such as feeding and spawning. By aligning fishing activities with these patterns, fishers can maximise their catch efficiency and minimise the environmental impact of fishing operations. Additionally, the scheduling of aquaculture activities, such as the harvesting of shellfish, often depends on tidal conditions to ensure the quality and safety of the produce.

Tourism, another significant economic sector, is also influenced by the tides. Coastal and marine tourism activities, including beach recreation, surfing, and wildlife viewing, are often dependent on tidal conditions. Beaches may be more accessible and attractive during low tides, while certain water sports are best enjoyed at high tides. Businesses catering to tourists must therefore consider tidal patterns in their operations, from scheduling tours to ensuring the safety of participants. Moreover, the aesthetic and educational value of tidal phenomena, such as tidal bores or the intertidal zones revealed at low tide, can attract visitors, contributing to the economic vitality of coastal communities.

Then there is tidal power, a form of hydropower that harnesses the energy generated by the natural rise and fall of ocean tides to produce electricity. This renewable energy source is particularly appealing due to its predictability and minimal environmental impact compared to fossil fuels. Tidal power plants operate by capturing the kinetic energy of moving water caused by tidal movements and converting it into electrical energy, typically through the use of turbines or barrages. Notable locations where tidal power is being harnessed include the Rance Tidal Power Station in France, the world's first and one of the largest tidal power plants, which has been operational since 1966. The Sihwa Lake Tidal Power Station in South Korea, another significant facility, utilises a seawall and turbines to generate electricity, making it the largest tidal power installation based on installed capacity. In the United Kingdom, the MeyGen project in Scotland represents one of the latest advancements in tidal stream technology, aiming to become the world's largest tidal stream project upon completion.

How does our moon's gravitational pull stack up with others?

The moon's tidal forces are not not unique to Earth and tides are not restricted to affecting only water. Across our solar system, other planets that host moons and satellites experience their own versions of tidal forces, which can have profound effects on the planet itself and any surrounding bodies. By comparing these tidal interactions within our solar system, we gain insights into the diverse outcomes of gravitational forces at play among various celestial configurations.

Jupiter, the largest planet in our solar system, offers a fascinating case study with its multitude of moons, the four largest being the Galilean moons: Io, Europa, Ganymede, and Callisto. The intense gravitational pull between Jupiter and these moons, particularly Io, which is the closest to the planet, leads to significant tidal heating. This process occurs as the varying gravitational forces exerted on Io by Jupiter and the other moons cause the moon to flex and distort. This flexing generates internal heat within Io, driving its intense volcanic activity, the most vigorous of any body in the solar system. Thus, while Earth's tides are most evident in the movement of water, Jupiter's gravitational influence manifests as geological activity on its moon, showcasing the variability of tidal effects.

Saturn and its moon Enceladus also illustrate the profound impact of tidal forces. Enceladus experiences tidal heating similar to Io, but with a crucial difference in its manifestation. The energy produced within Enceladus fuels geysers that eject water vapor and ice particles from cracks in the moon's icy surface, contributing to one of Saturn's rings. This interaction highlights how tidal forces can not only shape the geological features of moons but also contribute to the characteristics of the planets they orbit.

Neptune and its moon Triton offer another example of tidal dynamics, albeit with a twist. Triton orbits Neptune in a retrograde direction, opposite to the planet's rotation. This unusual arrangement leads to tidal deceleration, gradually slowing Triton's orbit. The energy dissipated by this process contributes to heating the moon's interior. Over astronomical timescales, this interaction is expected to lead to dramatic outcomes, potentially pulling Triton apart or causing it to crash into Neptune.

Mars, though it has two small moons, Phobos and Deimos, experiences tidal effects on a much subtler scale compared to the gas giants and their moons. Phobos, the closer of the two moons, is gradually spiraling inward toward Mars, a process driven by tidal interactions. This inward spiral means that Phobos will either break apart due to Mars's gravitational pull, forming a ring system, or crash into the planet in a spectacular impact event millions of years from now.

The study of tides and gravitational pull within our solar system reveals a complex tapestry of interactions that govern the behaviours and evolution of celestial bodies. From the volcanic eruptions on Io to the icy geysers of Enceladus and the eventual fate of Phobos, the impact of tidal forces extends far beyond the shores of Earth, playing a critical role in shaping the dynamics and destinies of planets and moons alike. But here on Earth, we have a very big and very close single moon, causing tides within our large blue ocean system. The moon and the sea together! Making love somewhere on a secluded beach. A poet's wet dream...

Sea-Fever
by John Masefield

I must go down to the seas again, to the lonely sea and the sky,
And all I ask is a tall ship and a star to steer her by;
And the wheel's kick and the wind's song and the white sail's shaking,
And a grey mist on the sea's face, and a grey dawn breaking.

I must go down to the seas again, for the call of the running tide
Is a wild call and a clear call that may not be denied;
And all I ask is a windy day with the white clouds flying,
And the flung spray and the blown spume, and the sea-gulls crying.

I must go down to the seas again, to the vagrant gypsy life,
To the gull's way and the whale's way where the wind's like a whetted knife;
And all I ask is a merry yarn from a laughing fellow-rover,
And quiet sleep and a sweet dream when the long trick's over.

CHAPTER 9

The new space race

The original space race was not only a catchy phrase flung about in the media in the 1960s and '70s, but a redirection of some potentially destructive technological advances that made the Cold War era... cold. Thankfully this era (1947–1991) never turned into a full scale global hot war. The Soviet Union and United States fought proxy wars in a few unfortunate countries around the world, but luckily for all humanity, the huge arsenal of nuclear devices they both built up were not used. Instead, scientific minds and bomb builders from both nations were redirected to the moon.

In today's geopolitical theatre, the moon has taken on a similar role to those heady times. Its craters and plains are becoming a new arena for international power dynamics, much like they were during the Cold War. The old space race was more than a quest for scientific knowledge; it was a demonstration of national strength and technological capability. However, unlike the two-nation tussle of the twentieth century, the contemporary lunar landscape is crowded with a diverse group of players, each with their own ambitions and aspirations.

The moon's strategic significance has evolved. What was once a finish line in a race for space supremacy has now become a potential launchpad for further space ventures and a source of valuable resources. The shift from a bipolar contest to a multipolar challenge has introduced a new complexity to space politics. Nations old and new to space exploration, along with private companies and tech billionaires, all see the moon as a critical stepping stone to broader cosmic ambitions and a canvas for technological display.

As we prepare to witness the next phase of lunar exploration, it's clear that the moon is more than just a celestial body—it's a mirror reflecting Earth's political rivalries and alliances. This rekindled lunar intrigue is not just about planting flags but about establishing a sustainable presence that could fuel future economies and technologies. The moon is becoming a testbed for international cooperation and competition, highlighting not only the advancements in space travel but also the ever-present human tendency to reach for the stars, both literally and metaphorically. Or to be more negative about the moon reflecting humanity—it might just be just a useless pile of dust. Surely not.

The moon as a strategic outpost

The geopolitical importance of the moon first became apparent during the Cold War's space race when the United States and the Soviet Union were eager to establish their dominance in space exploration. The Soviet Union's launch of *Sputnik* in 1957, the first artificial satellite, marked the start of a space competition that was as much about demonstrating military and technological might as it was about scientific achievement. Americans nervously watched the satellite circling the globe and tuned in to its ominous beep. It didn't seem very long before the Soviets put the first man into space in 1961.

That same year, President John F. Kennedy stood before Congress and ignited the ambition of a nation with an ambitious declaration that would define an era: the United States would send an American to the moon before the decade's end. This wasn't just a statement of intent; it galvanised the collective efforts of a country reeling from the Soviet Union's early successes in space. Kennedy's vision was clear and compelling, a rally to American innovation and resolve in the face of a daunting space race.

He framed the lunar mission not just as a technological challenge, but as a crucial front in the battle for global prestige and leadership during the Cold War. His speech would become a landmark moment, etching the moon mission into the national consciousness and setting the stage for what would become one of humanity's greatest achievements. The show was on, and ultimately broadcast to a television audience of 650 million people all around the world. Landing astronauts on the moon with the *Apollo 11* mission in 1969 was a giant technological and ideological victory. It was a moment that the U.S. could crow about for quite a while. There was now a new President Nixon making all the noise. For the moon was a bright light in a somewhat dark period for the United States: Kennedy assassinated, the war in Vietnam, rioting and a general distrust of government growing all the time.

However, the race to the moon was not solely about competition. It also led to moments of cooperation, like the Apollo–Soyuz Test Project, which showed that space could also be a place for peaceful collaboration. Back in the present day, with a variety of countries and private companies aiming for the moon, the landscape is quite different from the Cold War era. Enter the new tech billionaires on a stage with a shifting international power dynamic. This time, there's a more complex mix

of motivations driving the race, including national pride, scientific discovery, and commercial potential.

Military and strategic interests

The moon's role in military strategy has always been about the high ground, and in space, this concept takes on a literal dimension. It's very high ground. Establishing a presence on the moon offers a vantage point not just for observation but also for potential control over Earth's orbits. From the moment the Soviet Union's *Luna 2* mission impacted the lunar surface in 1959, signalling the capability to reach another celestial body, military minds have pondered the possibilities of lunar military assets. A lunar base could theoretically serve as a surveillance outpost, offering a clear, unobstructed view back to Earth and into the depths of space, potentially providing insights into satellite traffic and other strategic movements.

The idea of the moon as a staging ground for deep-space missions has evolved with our technological capabilities. In the wake of the Apollo program, which concluded in 1972, strategic focus shifted back to Earth. However, as countries have advanced their space technology and expanded their military aspirations to match, the moon's potential as a launch site for operations deeper into the solar system has regained interest. The concept is not just about exploration; it's about establishing a foothold in space from which to project power, whether for defence or for further ventures to other planets and asteroids.

In 1983, President Ronald Reagan introduced the Strategic Defense Initiative, colloquially known as "Star Wars," a futuristic plan aiming to protect the United States from nuclear attacks through a network of ground-based and space-based systems. The ambitious vision, part of a broader strategy to counter the Soviet missile threat during the Cold War, proposed using lasers, particle beams, and ground and space-based missile defence systems to intercept incoming warheads. Reagan's plan was as controversial as it was revolutionary, sparking debate over its feasibility, the potential militarisation of space, and the implications for the arms race. While the Strategic Defense Initiative was never fully developed or deployed, it marked a significant moment in the history of military strategy and space policy, reflecting the enduring theme of high-tech defence measures in the pursuit of national security.

Military interest in the moon was reignited in the twenty-first century, as countries like China landed *Chang'e 4* on the far side of the moon in 2019, underscoring the moon's potential as a strategic asset in space once again. The dual-use nature of space technologies means that advancements in navigation, communication, and propulsion for lunar missions also translate into military benefits.

Want to buy a block on the moon while you still can?

It's strange to think anyone has thought they could own any piece of the moon. However, people have paid money for what they think are their blocks on the moon—just look up the 'lunar registry' online and have your credit card ready.

As of 2024, the real estate prices are:

- *1 acre in the Sea of Tranquility – US$63.07 per acre*
- *3 acres in the Sea of Tranquility – US$56.76 per acre (SAVE 10%!)*
- *5 acres in the Sea of Tranquility – US$53.61 per acre (SAVE 15%!)*
- *10 acre estate in the Sea of Tranquility – US$50.46 per acre (SAVE 20%!)*

Internationally, there have been some more credible attempts to clarify this new frontier of real estate. The Outer Space Treaty of 1967 was the start. A foundational document ratified by spacefaring nations to govern the activities of states in the exploration and use of outer space. This treaty, with its principles of non-appropriation and peaceful use, effectively bars any nation from claiming sovereignty over celestial bodies, including the moon. Yet, as lunar missions become more ambitious and the prospect of mining lunar resources edges closer to reality, the treaty's broad stipulations are tested against new scenarios that its drafters could scarcely have envisioned.

The treaty's prohibition against national appropriation raises pressing questions about how resources extracted from the moon can be owned and used. As countries and private entities advance plans to harvest lunar water ice for life support and fuel, or mine rare minerals, the absence of a clear legal framework to regulate these activities invites uncertainty. This has spurred discussions among the international community to develop agreements that can fill these legal voids, ensuring that lunar resources can be utilised in a manner that benefits humanity while preserving the integrity and sustainability of the moon's environment.

The emergence of private actors in space exploration introduces additional layers of complexity. Companies investing in lunar missions seek assurances for their investments, prompting a re-examination of how space law accommodates private enterprises. The development of "safety zones" proposed by the Artemis Accords, for instance, is an attempt to balance the need for operational security with the provisions of the treaty. However, such initiatives also prompt debates over inclusivity and the equitable sharing of space benefits.

Over the years, there have been numerous instances where individuals and companies have claimed to sell plots of land on the moon, capitalising on the public's fascination with lunar ownership despite the clear legal stipulations set out by international treaties such as the Outer Space Treaty of 1967. These sales are often conducted through certificates of ownership for lunar land, marketed as novelty gifts or serious investments. However, these transactions exist in a legal grey area, as no entity on Earth has the authority to grant ownership of lunar land under international law. Critics argue that such sales are symbolic at best and misleading at worst, offering buyers a piece of paper rather than any enforceable rights or claims. Despite this, the practice has persisted, reflecting both the enduring human desire to connect with the cosmos and the complexities of applying Earth-bound legal concepts to the extraterrestrial realm.

The economics of lunar exploration: making real earth money

The economics of lunar exploration are complex to say the least. But that hasn't stopped anyone coming up with business ideas. At its core, the financial allure of the moon lies in its abundant resources and strategic position as a gateway for deeper space exploration. Realistically, making money from the moon involves tapping into its unique assets, from mineral wealth to its potential role in supporting future missions to Mars and beyond.

One of the most talked-about economic opportunities is lunar mining. The moon is thought to contain a wealth of valuable minerals and compounds, including rare earth elements crucial for electronics, renewable energy technologies, and helium-3, an isotope rare on Earth but abundant on the moon, which could theoretically be used in future nuclear fusion reactors. Extracting these resources could revolutionise industries by providing materials that are scarce or environmentally damaging to obtain on Earth. However, the technology for mining, processing, and transporting lunar resources efficiently and economically is still in development. The initial investment required is substantial, but the long-term payoff could be significant, not only in terms of raw materials but also by catalysing advancements in mining technologies and space infrastructure.

Another avenue for economic return is through the development of lunar infrastructure. Establishing a permanent human presence on the moon requires habitats, life support systems, energy generation, and communication networks. Companies specialising in aerospace, robotics, and construction could find lucrative contracts in designing and building these systems. Additionally, the moon could serve as a staging ground for missions further into the solar system, requiring services such as refuelling stations that use lunar water to produce rocket fuel. This infrastructure could also support scientific research, offering opportunities for companies to commercialise technologies developed for lunar conditions, ranging from advanced life support systems to new materials designed for extreme environments.

Tourism and branding are more speculative, yet potentially profitable, ways to make money from lunar exploration. As space travel becomes more accessible, the moon could become a destination for ultra-wealthy tourists seeking the ultimate off-world experience. Meanwhile, companies might leverage the moon in branding and marketing strategies, sponsoring missions or naming rights to lunar landmarks, blending the allure of space with commercial opportunities.

We will explore space tourism in the later chapter 'What it would be like to holiday on the Moon'.

International collaborations and tensions

Within the dynamic arena of lunar exploration, international collaborations and partnerships are emerging as pivotal elements, shaping the trajectory of humanity's return to the moon. The Artemis Accords, spearheaded by NASA, exemplify this new wave of cooperation, establishing a framework for space exploration that encourages participation from a global consortium of space agencies. These accords, which have seen signatories from countries as diverse as Britain, Japan, Australia, and the United Arab Emirates, aim to foster peaceful exploration, respect for heritage sites, and the transparent sharing of scientific data. Such initiatives underscore the moon's role as a platform for international diplomacy and collaboration, transcending terrestrial geopolitical boundaries.

However, these alliances are not without their complexities. They reflect the broader geopolitical landscapes and interests of the participating countries, intertwining scientific aspirations with national pride and strategic positioning. The collaboration on lunar projects brings together a wide array of technologies, expertise, and resources, necessitating a balance between cooperation and competition. As countries unite to tackle the technical and financial challenges of lunar exploration, they also navigate the delicate dance of international relations, ensuring that their contributions to joint missions enhance their standing and influence in both terrestrial and extraterrestrial spheres.

The emergence of these partnerships marks a significant shift in the governance of space activities. It heralds an era where multilateral agreements could pave the way for the sustainable and equitable use of lunar resources. It's a weird zone of diplomacy, that ultimately may help relations here on earth.

International space pals

As the moon becomes the next frontier for exploration, the dynamic between competitive nationalism and the spirit of global cooperation is akin to a high-stakes, interstellar tug of war. On one side, we have the age-old drive of nations to outdo each other, a kind of cosmic "keeping up with the Joneses" where the moon is the ultimate neighbourhood status symbol. Countries vie for the prestige of being first in various lunar achievements, from landing the next rover to establishing the first permanent base. This competitive zeal not only fuels rapid advancements in space technology but also stirs national pride and excitement back home. It's the space race 2.0, with a twenty-first-century twist, featuring a broader cast of characters eager to etch their names into lunar history.

There's a growing recognition that cooperation is in fact the only way to move forward. The challenges of returning to the moon, let alone setting up shop there, are immense. They're the kind of challenges that make you want to pool your resources, share your cosmic maps, and maybe even split the fuel bill. This cooperative approach is embodied in international treaties and partnerships, where knowledge, risks, and rewards are shared. It acknowledges that space is humanity's final frontier, not just the playground of any single country.

This balancing act between competition and cooperation adds a fascinating layer to lunar exploration. It's like a friendly neighbourhood rivalry that occasionally sees the competitors borrowing a cup of sugar or a lunar rover. As countries navigate their ambitions and the need for collaboration, they're writing the rulebook for the future of international space exploration. It's a unique blend of rivalry and partnership, pushing us towards a future where the moon might just be someone's next address.

This celestial diplomacy isn't just about carving up the moon pie; it's about ensuring that space remains a place for all humanity to explore and benefit from. It challenges us to think beyond national interests and work together for the collective good of humankind. As we take our first baby steps back on the moon, the hope is that this shared adventure sparks a new era of international cooperation, setting a precedent for how we tackle the even bigger challenges waiting for us among the stars. So, as we look up at the night sky, perhaps it's time to start seeing the moon not just as a distant satellite, but as the starting point for cooperation in the vast expanse of space. One day, could this giant dusty empty ball be full of happy space pals? The question is would those pals be human or something quite different?

Who owns the moon?

The evolving landscape of lunar exploration has seen commercial interests increasingly interwoven with scientific pursuits. This shift was highlighted earlier in 2024, when a US commercial spacecraft made an attempt to touch down on the lunar surface. The mission, conducted by Astrobotic Technology's Peregrine Mission 1, stirred controversy far beyond the realms of space enthusiasts and scientists. In the US, the Navajo Nation expressed their discontent, not with the scientific objectives, but with the more commercial aspects of the payload the lunar lander was carrying.

The Peregrine lander's cargo included 66 memorial capsules, which, for a fee of $US500 each, allowed individuals to send personal items to the moon, ranging from DNA samples to cremated remains. These included some of the cast and creators of *Star Trek*. This commercial offering, while seen by some as an innovative way to remember—if a slightly weird weird way to memorialise loved ones. This clashed with the Navajo people's cultural and spiritual values, which hold the

1. Russian space centre.
2. A SpaceX capsule.
3. The U.S.–Japan Space Collaboration Agreement.
4. Astronauts at work on the International Space Station. Is this what construction work on the moon will look like? 6. The International Space Station.

moon sacred. To them, the moon is not a distant celestial body to be used for commercial purposes but a sacred entity, integral to their cultural narratives and spiritual wellbeing.

But as has been the case for a few recent missions—the spacecraft didn't quite make it. The mission ended when a fuel leak forced a return to Earth, and its journey ended in a fiery re-entry rather than a historic lunar landing. The remains were burnt in space. This incident raises questions about the broader implications of commercial payloads in space missions. With more private companies entering the space arena, the line between scientific payloads and commercial ventures is becoming increasingly blurred. This trend underscores the need for a dialogue that reconciles the advancement of space exploration and commerce with the deep-rooted beliefs and cultural sanctity of communities like the Navajo people. As humanity reaches for the stars, it is confronted with the challenge of ensuring that space remains a realm that respects both our ambitions and our heritage.

While in this instance it was a commercial operator taking the human ashes to the moon, in 1999, the ashes of astrogeology founder Gene Shoemaker were sent there by NASA's Lunar Prospector space probe. At the time, the Navajo Nation lodged an objection against Shoemaker's ashes being taken to the moon and NASA then promised to consult in the future. This probe did land, and these ashes are still the only ones that have made it to the moon. But the laws are confusing up there and private companies are not bound by the same rules as NASA, for example. For the moment—if you can get there in your own rocket—you can do what you want.

CHAPTER 10

The mega rich in space

Imagine a world where blasting off into space is as easy as catching a flight to New York or Paris. Sounds like a script from a sci-fi movie, right? Well, hold onto your space helmets, because this is rapidly becoming our reality, thanks to a trio of billionaires with sky-high ambitions. Elon Musk, Jeff Bezos, and Richard Branson are not your average entrepreneurs; they're space wannabe pioneers on a mission to turn the cosmos into the next frontier for adventure, industry, and human expansion.

Gone are the days when space exploration was the exclusive playground of government agencies like NASA and Roscosmos. Enter the era of SpaceX, Blue Origin, and Virgin Galactic—companies that are making space more accessible and, dare we say, *cooler* than ever before, despite the fact these men come across as mega nerd bond villains. These aren't just venturers with their eyes on the stars; they're trailblazers crafting the future of space tourism, moon missions, and even the prospect of living on Mars.

This isn't just about rich guys having a blast in space… although… in reality, it actually is… It's a seismic shift in how we approach the great beyond. Space tourism? Check. Colonising the moon and beyond? On the to-do list. A burgeoning space economy that could redefine how we live, work, and play? Absolutely. And with this audacious leap comes a whole galaxy of questions about policy, regulation, and innovation. It's a wild, exhilarating time to be alive, as we stand on the brink of a new chapter in human history, one where the final frontier is closer than ever. So, strap in, folks. The journey into the future of space exploration, powered by some of the richest visionaries on (and off) the planet, is about to take off. It's like we are in a science fiction movie scene where the elite get to take off in a space ship while the earth explodes. Or is it the following scene where we see the fake exploding earth projected onto their spacecraft windows and we cut back to the next scene below, where all the good people rejoice on earth? Everyone happy and free, and the baddies gone for ever…

A new era of space exploration

The twenty-first century is a new phase that represents a significant departure from the traditional paradigms of space exploration, which were predominantly characterised by government-led missions undertaken by national space agencies such as NASA, Roscosmos, and ESA. The shift towards commercialisation has been both rapid and transformative, altering not only the landscape of space exploration but also the very framework within which future missions are conceived and executed.

At the heart of this transformation is the involvement of mega-rich entrepreneurs who have turned their sights towards the cosmos, driven by a blend of ambition, curiosity, self preservation and the desire to push the boundaries of what is possible. They are going about their various projects with a missionary zeal. This cadre of space pioneers is not content

THE MEGA RICH IN SPACE

with merely investing in space as an abstract concept; rather, they are actively developing the technologies and infrastructure needed to make space more accessible, and in doing so, they are redefining humanity's relationship with the final frontier. Elon Musk of Space X has his eyes on the moon as a bus stop for Mars.

> *"There needs to be an intersection of the set of people who wish to go, and the set of people who can afford to go... and that intersection of sets has to be enough to establish a self-sustaining civilisation. My rough guess is that for a half-million dollars, there are enough people that could afford to go and would want to go. But it's not going to be a vacation jaunt. It's going to be saving up all your money and selling all your stuff, like when people moved to the early American colonies... If we can establish a Mars colony, we can almost certainly colonise the whole solar system, because we'll have created a strong economic forcing function for the improvement of space travel."*

It's a pity Elon Musk has gone a bit Moonraker over time. Still, SpaceX continues as a major player in the new space race. The significance of this shift to a Wild West attitude to space exploration cannot be overstated. For the majority of history, the vast resources required for space missions meant that only national governments could afford the colossal investment in research, development, and execution of space missions. However, the entry of private players has introduced a new dynamism into the field, characterised by innovation, flexibility, and a willingness to take risks that traditional space agencies, bound by bureaucratic constraints and public accountability, could ill afford. For instance Musk's SpaceX will purposely

U.S. President Barack Obama with SpaceX entrepreneur, Elon Musk.

blow up rockets to test them. They call it 'disassembly'. Or perhaps they're just blowing up things for fun.

These entrepreneurs are not just focusing on satellite launches or short-term orbital flights. Their ambitions are far more expansive, targeting not only Earth orbit but also the moon, Mars, and even the broader solar system. Their vision encompasses a future where space travel is not a rarity but a routine, where space tourism is a thriving industry, and where humanity is taking its first tentative steps towards becoming a multi-planetary species.

This new era of space exploration is also distinguished by the collaborative and competitive spirit that these private ventures bring to the table. Competition among companies like SpaceX, Blue Origin, and Virgin Galactic drives technological advancements at a pace seldom seen in the traditionally slow-moving aerospace sector. This competition is paralleled by collaboration, both among the private entities and between the private and public sectors, creating a synergistic environment that accelerates progress towards shared objectives.

The involvement of these entrepreneurs in space exploration is catalysing interest and investment in related fields, from satellite technology and telecommunications to planetary science and astrobiology. The ripple effects of their ventures are not only confined to space travel alone but are fostering innovation across a broad spectrum of technologies, with potential benefits extending well beyond the aerospace industry.

The foray of these super rich, sometimes sinister entrepreneurs, into space exploration marks a critical moment in the history of humanity's quest to explore and understand the cosmos. By infusing the sector with capital, competition, and a vision that stretches to the far reaches of space, they are not only opening new frontiers for exploration but also laying the groundwork for a future in which space is an integral part of human life. It may just be because private companies can fire people so much more easily than government organisations that they are able to be flexible and innovative. Making for a somewhat terrified, compliant workforce. Whatever the reasons, this new era of space exploration, promises to be very interesting.

The prominent figures of the space exploration

At the forefront are three figures: Elon Musk, Jeff Bezos, and Richard Branson. They cast long shadows over the future of the moon. Dysfunctional Elon, buff billionaire Jeff, depressingly cheerful Richard. These entrepreneurs have transcended their earthly successes to become the pioneering leaders of a transformative period in space exploration, each driven by a unique blend of personal ambition, technological innovation, and a vision that extends their reach far beyond the familiar blue skies of earth.

Musk, the mind behind companies like Tesla, Inc. and PayPal, founded SpaceX with an audacious goal: to make space transportation affordable, thereby enabling the colonisation of Mars. Fueled by a conviction that humanity must become a multi-planetary species to ensure its survival, Musk has faced skepticism head-on. Through determination and a series of groundbreaking innovations in rocket design and manufacturing, SpaceX has shattered industry norms, achieving milestones such as launching the first privately funded, liquid-fueled rocket to reach orbit and sending the first privately funded spacecraft to the International Space Station. Musk's disruptive vision for space exploration has not only revolutionised the industry but also sparked a global renaissance in space enthusiasm and investment.

Bezos, the entrepreneur who revolutionised e-commerce with Amazon, has harboured a lifelong passion for space. Blue Origin, his foray into the final frontier, is founded on a vision of enabling millions to live and work in space, benefiting Earth with the resources and tranquility of the cosmos. Bezos's methodical, long-term approach is encapsulated in Blue Origin's motto, *gradatim ferociter*, or "step by step, ferociously." The company has focused on developing reusable rocket technology, aiming to lay the foundations for a future where space travel is as accessible and routine as air travel, with an emphasis on suborbital space tourism as a stepping stone towards grander ambitions.

> "The Moon Village concept has a nice property in that it basically just says, 'look, everybody builds their own lunar outpost, but let's do it close to each other.' That way... you can go over to the European Union lunar outpost and say, 'I'm out of eggs. What have you got?"
>
> *Jeff Bezos, Blue Origin*

Branson, the hyper cheerful British entrepreneur behind the Virgin Group, entered the space race with a different angle. Virgin Galactic aims to make space travel accessible to all, not just trained astronauts. By developing SpaceShipTwo, a spaceplane designed for suborbital space tourism, Branson seeks to democratise access to space, allowing people to experience the profound beauty and weightlessness of the cosmos firsthand. Branson's vision focuses on the human experience of space travel, aiming to inspire and broaden the horizons of those who venture beyond Earth's atmosphere.

> "Ever since I saw the moon landing as a young teenager, I was determined I would go into space one day."
>
> *Richard Branson*

The ventures led by Musk, Bezos, and Branson—SpaceX, Blue Origin, and Virgin Galactic— are the brightest lights of this new era of space exploration, emblematic of the shift towards commercial space travel. Each company, reflective of its founder's vision, contributes uniquely to the collective endeavour of pushing human presence beyond Earth. SpaceX's pioneering efforts in making interplanetary travel feasible, Blue Origin's dedication to sustainable space infrastructure, and Virgin Galactic's mission to open space to everyone highlight the diverse pathways through which private enterprise is contributing to humanity's celestial aspirations.

The technical achievements of this group point to a possible future where humanity's footprint in space extends to the moon, Mars, and beyond. As these companies innovate and challenge the status quo, they redefine what is possible, heralding the dawn of an era where space exploration is driven not solely by national agencies but by tech billionaires.

Together, Musk, Bezos, and Branson embody the spirit of the modern space age—a blend of bold vision, relentless innovation, and an unwavering belief in space's potential to inspire and transform humanity. Their contributions are charting the course for future generations in space, for better or worse.

Are these the right people to be involved in space exploration?

The burgeoning era of space travel has sparked a global conversation about the implications of the private sector's ascendancy in what was once a predominantly government-led domain. This shift raises several critical questions, particularly regarding the motivations of these capitalist titans, their track records in other industries, and the potential impact of their personal ideologies on the future of space exploration. Government organisations like NASA used to control space through their monopoly on the spare industry. Tech billionaires have used their monopolies to amass huge amounts of money, to challenge the old ways.

At the heart of this debate is the inherent tension between the profit-driven motives typical of capitalist ventures and the altruistic ethos that many believe should guide humanity's journey into space. The concern is not just theoretical; it is rooted in tangible issues observed within these entrepreneurs' terrestrial businesses. For example, Amazon, under Bezos, has faced criticism for workplace conditions and pay scales that many argue do not reflect the vast wealth of the company or its founder. Similarly, Musk's activities on social media platforms, including his acquisition and alleged transformation of Twitter, have sparked discussions about his broader agendas and how they might influence his space exploration endeavours.

Critics argue that if space travel and exploration are primarily controlled by individuals whose businesses have been criticised for prioritising profits over people, there might be cause for concern. The fear is that space could become a playground for the wealthy, with access to space travel and its benefits skewed towards those who can afford it, potentially sidelining the scientific, exploratory, and humanitarian goals that have traditionally driven space exploration. The personal ideologies of these entrepreneurs, if reflected in their space ventures, could colour the priorities of humanity's presence in space, influencing everything from the selection of missions to the cultures and values propagated beyond Earth.

> "There's a silly notion that failure's not an option at NASA. Failure is an option here. If things are not failing, you are not innovating enough."
>
> Elon Musk

However, it's essential to consider the other side of the coin. The involvement of Musk, Bezos, and Branson in space travel has undeniably accelerated progress in the field. Their companies have introduced groundbreaking technologies, reduced the cost of access to space, and revitalised global interest in space exploration. The competitive dynamics fostered by their participation have spurred innovation at a pace that traditional, government-led programs might not have achieved on their own. Their vision for space—ranging from Mars colonisation to space tourism—has expanded the scope of what many considered possible, opening up new avenues for exploration, settlement, and even economic development in space.

The crux of the matter lies in finding a balance. The motivations of these entrepreneurs, rooted in a capitalist framework, are not inherently at odds with the broader goals of space exploration. However, the transition towards a private sector-dominated era of space travel necessitates robust regulatory frameworks and international cooperation to ensure that space remains a realm for the benefit of all humanity, not just the elite. Policies and agreements must be crafted to safeguard ethical standards, ensure equitable access, and preserve space as a shared resource for scientific discovery and exploration.

Having gained such wealth and power, these individuals are not always well loved by everyone. Becoming mega rich is now synonymous with disruption. But these people are the ones with the money and will to make things happen out there.

Beyond the orbit of the Earth

The significance of targeting the moon lies not just in the scientific and exploratory value such missions entail but also in the strategic importance of establishing a human presence there. The lunar surface offers invaluable resources and serves as a proving ground for technologies essential for longer-duration space travel, such as life support systems, habitat construction, and in-situ resource utilisation. The moon also provides a strategic platform for launching missions further into the solar system, including Mars and beyond.

SpaceX is actively working towards enabling future lunar exploration through its development of the Starship spacecraft. While initially focused on Mars, SpaceX's architecture is versatile enough to facilitate missions to the Moon. The company has been selected by NASA to develop a lunar version of Starship to land astronauts on the moon as part of the Artemis program, marking the first time a private company has been tasked with such a mission. This collaboration underscores the critical role that SpaceX's innovation and technology play in advancing the frontier of lunar exploration.

Blue Origin also has its sights set on the moon, with its Blue Moon lander project. Designed to deliver a variety of payloads to the lunar surface, including rovers, scientific instruments, and eventually humans, Blue Moon embodies Blue Origin's commitment to supporting a sustained human presence on the Moon. Bezos envisions the moon as a crucial location for off-Earth manufacturing and resource extraction, leveraging the lunar resources to benefit Earth and facilitate further space exploration.

Virgin Galactic, while primarily focused on suborbital space tourism, contributes indirectly to the goal of lunar and beyond Earth orbit exploration by democratising access to space. By making space more accessible, Virgin Galactic is helping to foster a broader public interest and investment in space exploration. This growing interest could support future missions to the moon and beyond by generating a wider base of support for space exploration efforts.

The endeavours of these companies and their leaders represent a concerted effort to extend humanity's reach to the moon and establish a sustainable presence there, laying the groundwork for the future exploration of Mars and other destinations in the solar system. They see worth in pursuing in this push for the moon. Whatever it is, however, is sometimes hard to dig out. Bezos believes there's something worth mining there, with all the sunlight hitting the moon all the time—at least one side of it—and he seems to be suggesting it will benefit the Earth's environment.

> *"The Earth is not a very good place to do heavy industry. It's convenient for us right now. But in the not-too-distant future—I'm talking decades, maybe 100 years—it'll start to be easier to do a lot of the things that we currently do on Earth in space, because we'll have so much energy. We will have to leave this planet. We're going to leave it, and it's going to make this planet better."*
>
> Jeff Bezos

One catch though: it's very hard to get there and even harder to bring anything back.

The shift from government to private sector

The monumental achievements of the mid-twentieth century, most notably the Apollo moon landings, were the result of massive, state-funded programs undertaken by superpowers in the context of the Cold War. These endeavours were driven by a combination of geopolitical competition, scientific curiosity, and national pride. The immense costs, technological challenges, and high risks associated with space exploration meant that only the resources of nation-states could support such ambitious projects.

Why is it starting to look a lot better for private companies? First, advancements in technology have significantly reduced the cost of accessing space, making it more feasible for private entities to develop and launch spacecraft. Second, the commercial potential of space—from satellite communications to space tourism and beyond—has attracted investment and innovation from the private sector. Finally, the dogged vision of these new entrepreneurs has lit a fuse under the rocket industry.

The shift towards private sector involvement in space exploration has been both encouraged and supported by government agencies. For instance, NASA has played a pivotal role in fostering this new ecosystem through initiatives such as the Commercial Orbital Transportation Services and the Commercial Crew Program. By contracting out cargo and crew transport to the International Space Station to companies like SpaceX and Orbital ATK (now part of Northrop Grumman), NASA has not only stimulated the commercial space sector but also focused its own resources on deeper space exploration missions, such as the Artemis program to return humans to the moon.

Private companies, with their ability to innovate rapidly and operate flexibly, have introduced new technologies and reduced costs through efficiencies and economies of scale that were previously unattainable. The competition among private entities has spurred innovation and accelerated the development of new spacecraft, launch systems, and related technologies. The private sector's involvement has expanded the scope of space exploration to include not only scientific and exploratory missions but also commercial ventures, thereby broadening the base of support for space activities.

This transition from government to private sector-driven space exploration represents a paradigm shift in how humanity approaches the final frontier. It signals a move towards a more collaborative model, where government agencies and private companies work together to achieve shared goals. This model leverages the strengths of each sector: the experience, infrastructure, and regulatory capabilities of government agencies, and the innovation, agility, and funding of the private sector. What could possibly go wrong?

Space tourism

The concept of space tourism, once relegated to the realm of science fiction, is rapidly evolving into a viable industry, marking a significant and exciting development in the broader context of space exploration. This burgeoning sector is a direct result of the transformative shift in space exploration from a government-dominated endeavour to one that includes substantial participation by the private sector. As commercial enterprises have begun to play a more prominent role in space travel, the dream of ordinary people visiting space is inching closer to reality.

The genesis of space tourism can be traced back to the early 2000s when the first privately funded space tourists visited the International Space Station aboard Russian Soyuz spacecraft. However, it is the recent advancements and investments by private space companies that have truly ignited the potential for space tourism to become a mainstream activity. Companies like Virgin Galactic, Blue Origin, and SpaceX are at the forefront of this revolution, each offering a unique approach to making space more accessible to the public.

Virgin Galactic is focusing on suborbital flights that offer several minutes of weightlessness and breathtaking views of Earth from the edge of space. Their spaceplane, *VSS Unity*, has successfully reached space, demonstrating the feasibility of commercial suborbital space tourism. Blue Origin's *New Shepard*, a rocket that launches vertically and returns to land on Earth, offers a similar suborbital experience, with the capsule providing large windows for unparalleled views of space. Meanwhile, SpaceX has set its sights higher, planning not only suborbital but also orbital flights, including proposed missions around the moon, which would offer civilians the chance to experience deeper space travel.

The implications of the development of space tourism extend far beyond the opportunity for individuals to experience space firsthand. Economically, the space tourism industry has the potential to become a significant market segment, generating revenue that can be reinvested into further space exploration and technology development. This reinvestment can accelerate advancements in spacecraft design, propulsion systems, safety measures, and sustainability practices, benefiting all aspects of space exploration.

Space tourism has profound implications for society and culture. By democratising access to space, it has the potential to change the way people view our planet and humanity's place in the universe. The overview effect, a cognitive shift in awareness reported by astronauts who have seen Earth from space, emphasising its fragility and the need for stewardship, could influence a broader segment of the population. This heightened collective consciousness about our planet could catalyse greater interest in protecting our environment and promoting global unity.

However, the future of space tourism is not without challenges. Issues such as the environmental impact of rocket launches, the safety of passengers, regulatory frameworks, and the high cost of space travel are all hurdles that need to be addressed. As the industry matures, it will be crucial for companies to tackle these challenges head-on, ensuring that space tourism develops in a sustainable, safe, and equitable manner.

In the coming years, space tourism is poised to expand dramatically, with more companies entering the market and technological advancements making space travel safer and more affordable. While the initial phase of space tourism may be accessible only to the wealthy, the long-term vision is much more inclusive, aiming for a future where space travel is an option for a broad spectrum of society.

Space colonisation on the moon and other planets

The concept of space colonisation, particularly on the moon and other planets, represents one of the most ambitious and visionary goals of the current era of space exploration. This aspiration goes far beyond the immediate ambitions of space tourism, aiming to establish permanent human settlements beyond Earth. This vision is not merely an extension of human exploratory spirit but a pragmatic step toward ensuring the long-term survival and prosperity of humanity.

Space colonisation involves the development of sustainable habitats where humans can live and work for extended periods. The moon, with its proximity to Earth, serves as an ideal stepping stone for the colonisation of deeper space, offering a platform for scientific research, technological development, and as a base for missions further into the solar system. Mars, with its more Earth-like characteristics compared to other celestial bodies, has been identified as a prime candidate for colonisation, with its potential for water, essential minerals, and the possibility of terraforming to create a more Earth-like environment.

> *"It's important to get a self-sustaining base on Mars because it's far enough away from earth that in the event of a war, it's more likely to survive than a moon base. If there's a third world war, we want to make sure there's enough of a seed of human civilisation somewhere else to bring it back and shorten the length of the dark ages."*
>
> Elon Musk

We can clearly see what Mars is like via images transmitted from the rovers that have been successfully sent grinding around the planet in the recent past. It still doesn't appear to be a place you'd want to raise your kids, yet the dream lurches on. What would *we* get out of it?

Scientifically, establishing colonies on the moon and Mars would provide opportunities for research in a wide range of fields, from astrobiology and geology to human physiology in reduced gravity environments. Economically, the extraction of resources such as rare minerals and Helium-3 (a potential future energy source) from the lunar surface could offer significant benefits. Strategically, space colonies could ensure humanity's survival in the event of catastrophic events on Earth, such as asteroid impacts, nuclear war, or environmental collapse.

At this very moment, a range of companies are developing technologies critical to long-term habitation in space, including life support systems, habitats capable of protecting inhabitants from space radiation and micrometeorites, and technologies for in-situ resource utilisation to extract water and produce fuel from local resources. The collaboration between private enterprises and government agencies is crucial in this endeavour, combining resources, expertise, and regulatory frameworks to pave the way for future colonies.

However, the challenges of space colonisation are immense and multifaceted. Technologically, the hurdles include creating reliable life support systems, ensuring the psychological and physical health of colonists, and developing efficient transportation systems for traveling between Earth and the colonies. Ethically and legally, questions about the governance of extraterrestrial colonies, the protection of celestial bodies from contamination, and the rights of colonists must be addressed. Even though many of the potential locations look pretty gnarly, the impact of human activity on pristine celestial environments needs careful consideration to avoid the mistakes made on Earth.

Despite these challenges, the progress in space technology and the growing interest in space exploration suggest that space colonisation could become a reality within the next few decades. Initial steps might involve the establishment of small, permanent research bases on the moon, followed by larger habitats capable of supporting communities of scientists, engineers, and eventually families. Over time, as technologies advance and experience is gained, these outposts could evolve into self-sustaining colonies, marking the beginning of a multi-planetary human civilisation.

1. Jeff Bezos' Blue Origin facility. 2. Virgin Altantic media conference at NASA. 3. The always-effervescent Richard Branson. 4. Elon Musk with the powers-that-be at NASA. 5. Three astronauts on the moon, but can Bezos, Branson and Musk work together? 6. Elon Musk at SpaceX.

The future and development of the space economy

A space economy seems a far-fetched concept at this point. To build something profitable from something so incredibly expensive seems difficult to say the least. But money is a totally human construct, and so making money is in itself a leap of faith. The mega rich space team have built their own wealth in ways that are beyond the understanding of most people. It's often imaginary money made of debt, swirling around the globe, making economic machines swing into action. Somehow, debt makes them rich.

This new wave of space exploration, driven by both the rich and government agencies, is laying the groundwork for what promises to be one of the most transformative developments in human history: the emergence of a space economy. This nascent economy extends the economic sphere of human activity beyond Earth, opening up new frontiers for commerce, science, and industry.

The concept of a space economy is not merely about extending current economic activities into space but involves the creation of entirely new markets and opportunities. Space tourism, as discussed, is just the tip of the iceberg. Beyond the allure of space travel for private individuals lies the vast potential for resource extraction from asteroids, the moon, and other celestial bodies. These resources range from water, which can be used in life support systems or split into hydrogen and oxygen for rocket fuel, to rare minerals and metals that are scarce and expensive on Earth. The economic implications of tapping into these resources could revolutionise industries such as energy, construction, and electronics.

Another pillar of the future space economy is manufacturing in space. The unique conditions of microgravity offer opportunities for research and production that are impossible or highly inefficient on Earth. For instance, the production of certain types of fibre optics, pharmaceuticals, and materials can benefit from the lack of gravity, leading to products of higher quality and performance. As transportation costs to and from space decrease and in-orbit manufacturing technologies advance, the economic viability of space-based manufacturing is expected to increase significantly.

The development of habitats and infrastructure in space is also a critical component of the space economy. As humans begin to spend more time in space, whether for research, tourism, or as part of colonisation efforts, there will be a growing need for the construction of living and working spaces. This includes not only habitats on celestial bodies such as the moon and Mars but also space stations and habitats in orbit. The construction and maintenance of these facilities will require a wide range of services and supply chains, from the provision of life support systems to the delivery of food and other essentials. Should we call the people who will do these jobs 'spacies'? High-vis spacesuits on the way.

Policy and regulation implications for moon and space exploration

In the Wild West depicted in American mythology, and made global via Hollywood movies, there was always a point where the sheriff came into town, bringing the rule of law, after a big shootout. One of the primary implications of an extraterrestrial economy is the need for clear and comprehensive policies regarding the use of space and celestial bodies. The Outer Space Treaty of 1967, which forms the basis of international space law, establishes that space shall be free for exploration and use by all countries and prohibits national appropriation by claim of sovereignty, by means of use or occupation, or by any other means. However, the treaty was crafted in an era dominated by government-led space exploration, and its provisions are broad and open to interpretation. As commercial entities begin to play a larger role, there is a pressing need for additional treaties, agreements, and national legislation that address the specifics of commercial exploitation, resource extraction, and the establishment of habitats or colonies on celestial bodies.

Safety and liability are also critical areas that require updated policies and regulations. The increasing frequency of commercial launches and the emergence of space tourism raise questions about the safety standards that private space companies must adhere to, the certification of spacecraft and crew, and the liability in the event of accidents. International agreements and national laws need to clearly delineate the responsibilities of commercial space operators, ensuring the safety of crew, passengers, and cargo, as well as the protection of Earth's environment from potential hazards such as space debris or contamination from extraterrestrial sources.

As commercial activities expand to include resource extraction from asteroids, the moon, or other planets, it is essential to establish guidelines that prevent the over-exploitation or harmful alteration of these celestial bodies. Policies must encourage the sustainable and responsible use of space resources, including measures to prevent the creation of space debris and to ensure the preservation of historically or scientifically significant sites.

The equitable access to space and its benefits is another significant consideration. The advent of commercial space activities poses the risk of exacerbating inequalities between nations and within societies, with wealthier countries and individuals having disproportionate access to space resources and opportunities. International cooperation and agreements are needed to ensure that the benefits of space exploration and the space economy are shared equitably, promoting the peaceful use of space and fostering international collaboration.

The regulation of commercial space activities must also address issues of national security and the potential militarisation of space. Policies and agreements should aim to prevent an arms race in space, ensuring that space remains a domain for peaceful activities and scientific exploration. This includes regulating the deployment of weapons in space, the use of satellites for military surveillance, and other activities that could threaten global security.

It's that point in a Wild West movie, where the sheriff gets an office and sits down to write some tedious documents. A scene not often included in the average Western.

Technological innovation

The relentless pursuit of space exploration by this posse of spacemen is not just reshaping the narrative of humanity's quest beyond Earth; it is also a catalyst for technological innovation in space travel and exploration. The infusion of competition and significant private investment into the space sector by figures such as Musk, Bezos, and Branson has ignited a renaissance of innovation, propelling the development of new technologies and methodologies that promise to redefine our approach to exploring the cosmos.

The dynamics of competition in the private space industry have proven to be a powerful force for innovation. Unlike traditional government-led space programs, which often operate under the constraints of budget allocations and political oversight, private companies can move with greater flexibility and a higher tolerance for risk. This agility enables them to pursue ambitious projects and adopt cutting-edge technologies that push the boundaries of current capabilities. For instance, the quest to reduce the cost of space travel has led to the development of reusable rocket technology, a concept that was once deemed impractical by many in the aerospace community. Companies like SpaceX have successfully demonstrated the feasibility and efficiency of reusable rockets with their Falcon series, significantly lowering the cost barrier to space and enabling a higher frequency of launches.

In addition, the personal investment and involvement of these entrepreneurs in their space ventures infuse their companies with a sense of mission and urgency that drives innovation. Their willingness to invest vast resources into research and development has accelerated the pace of technological advancement in the sector. For example, the development of powerful and efficient rocket engines, lightweight materials for spacecraft, and sophisticated life support systems have all benefited from the increased investment flowing into the space industry.

The competitive landscape has also fostered a culture of innovation that extends beyond individual companies. Collaborations between private companies and government agencies, such as NASA's partnerships with SpaceX and Boeing to transport astronauts to the International Space Station, leverage the strengths of both the private and public sectors. These partnerships not only facilitate the sharing of knowledge and resources but also stimulate the development of standards and practices that further innovation in the industry.

The ambitions of these space entrepreneurs extend to the development of technologies that could support life in space for extended periods, including habitats on other planets.

This long-term vision necessitates innovations in a wide array of fields, from bioregenerative life support systems to technologies for in-situ resource utilisation, which would allow future explorers and settlers to utilise local materials for construction, fuel, and other necessities.

The impact of technological innovation driven by competition and investment in the space sector extends far beyond the immediate applications in space travel and exploration. Many technologies developed for space have found applications on Earth, improving products and services in industries such as telecommunications, manufacturing, healthcare, and transportation. This phenomenon, known as technology transfer, underscores the broader benefits of investing in space exploration.

The involvement of the billionaire gang in the space sector has been a significant force for technological innovation, challenging conventional wisdom and inspiring a new generation of engineers, scientists, and dreamers. But the big question is who will get there first to plant the corporate flag? Our bet is that it will be a big 'X'. Branson seems to be looking for a joy ride fairly close to Earth with Virgin Galactic. There is a history of conflict between the other two contenders, Space X and Blue Origin, so that race will be very interesting. Musk comes across as the most moon-driven, although Bezos with his interest in life extension may outlive any competition by the time such an event occurs.

CHAPTER 11

A holiday on the moon: is it really possible?

When this writer was a child in the 1970s, we were given the hard sell on a holiday on the moon. There were children's books like *You Will Go To The Moon*; Tintin was up there frequently. Thunderbirds were go, and *My Favourite Martian* flew past it on his way to Earth. It seemed like it would happen pretty soon, and if you look through more serious predictions from that period, it was the same message. Futurologists were sure that once 'man' was on the moon, others would quickly follow. A Hilton Hotel was even planned for the moon.

> *'The best way to stay ahead in the hotel business is to stay ahead of other people in the hotel business...'*

The Lunar Hilton was going to be a real thing. At least in the minds of the Hilton company and their advertising people. It represented a fascinating intersection between the golden era of space exploration and the boundless optimism of commercial enterprise. First conceptualised amidst the space-themed euphoria of a 1958 company event in Chicago, the idea gained real momentum in the imagination of Barron Hilton during the 1960s. As humanity stood on the cusp of actualising its lunar ambitions, Hilton envisaged not just planting flags on the dusty lunar surface but constructing a haven of hospitality—a hotel that would defy every traditional notion by offering accommodation in the most alien of environments. His proposal for a 100-room underground hotel, complete with a 24-room space station orbiting above, was a bold statement of faith in human progress and ingenuity. It was a vision far beyond its time, promising a blend of luxury and adventure that was quintessentially Hilton, yet undeniably futuristic.

The Lunar Hilton concept was more than just architectural blueprints and financial projections; it was a masterstroke of marketing that captured the *zeitgeist* of the space age. Potential guests were sent reservation cards and mock room keys. This was at a time when the Apollo missions were not just breaking new ground but also shaping the collective consciousness regarding space. Hilton's idea tapped into the excitement of the times, promising a future where space tourism was not a mere fantasy but an impending reality. The Lunar Hilton, became a symbol of optimism, a beacon that reflected humanity's longing to reach the stars and the belief that where we go, our comforts and luxuries could follow.

The resurgence of interest in the Lunar Hilton during the 1990s demonstrated the enduring appeal of the concept. As the world looked back at the moon with renewed interest and ambition, the plans for this extraterrestrial hotel grew more grandiose. The updated vision for a 5,000-room hotel was ambitious. This iteration, drawing upon the technology of the era, proposed utilising recycled Space Shuttle fuel tanks to construct a space station—an idea that meshed sustainability with innovation. Although the Lunar Hilton remains a concept yet to be realised it's a concept that refuses to go away.

Lunar holiday

The lunar holiday seemed tangible In the heady days following the Apollo landings. The idea of sipping a cosmic cocktail while gazing at the Earth hanging in the star-studded sky seemed just around the corner. The attractions of a lunar holiday would be out of this world—literally. Zero-gravity sports where you could somersault like an Olympian, the thrill of bounding over the moon's desolate beauty in a lunar rover, or the chance to leave your footprints next to those of Armstrong and Aldrin.

The 1970s vision of moon tourism was one of cosmic optimism. Imagine lunar lodges with viewing domes where Earthrise and sunset would occur several times a day, offering a panorama of our blue planet rising over the desolate, yet serene, lunar horizon. Entertainment would not just be a sideshow; it would be an experience heightened by the stark and majestic moonscape. From space-jazz festivals to moonwalk dance-offs, the entertainment industry would have a field day—or rather, a field orbit—in crafting experiences that defy gravity in every sense.

The moon also offered tranquillity, perhaps inspired by the lunar 'sea' of that name. In the 1970s, it offered a retreat where the silence of space envelops the guest, a stark contrast to the hustle and bustle of Earthly life. Yes, they considered urban life stressful in the 1970s, just like we do today.

The moon would be a place for reflection, a unique vantage point to ponder our place in the cosmos. The dreamers of the 1970s foresaw that the moon could offer not just a physical journey, but a spiritual one too, where looking back at Earth could turn a holiday into a life-changing perspective shift.

Lunar tourist attractions

The allure of a lunar holiday lies in its promise of once-in-a-lifetime experiences, set against the backdrop of the moon's stark and otherworldly beauty. Picture this: tourists partaking in low gravity sports, a spectacle where each leap is a flight and each step defies the very essence of weight as known on Earth. Activities might range from lunar golf, where a swing sends the ball soaring over the horizon, to volleyball, where serves and spikes become graceful arcs in slow motion. The reduced gravity would not only redefine sports but also open the door to entirely new recreational activities, ones that we can hardly imagine from our terrestrial perspective.

Then, of course, there is sex. Low gravity sex could either be very exciting or very frustrating. NASA claims that no humans have yet had sex in space. Data from the Russians is less clear, but it's an obvious area that holidaying humans would be keen to explore. The 'mile high club' would expand to the roughly '238,855 mile high club'. Sex made the internet, why not space tourism? That knowing glance from your fellow moon traveller says everything—no need for a moonlit serenade; you're on it already. And you both took the trouble to earn a lot of money and make this very expensive trip. What could be more romantic?

Then there are the vistas that would take your breath away—quite literally, if not for the life-support systems in your habitat. The Earth views from the moon are not merely sights to behold; they are perspectives that redefine our place in the universe. The iconic Earthrise, first witnessed by the crew of *Apollo 8*, would be an everyday spectacle, serving as a humbling and beautiful reminder of our home planet's fragility in the vast expanse of space. Tourists could enjoy these views from the comfort of their domed accommodations or during a serene stroll on the lunar surface.

Finally, what lunar holiday would be complete without a touch of historical reverence? Tours of the Apollo landing sites would be akin to walking through open-air museums dedicated to human tenacity. Each footprint and left-behind artefact tells a story of the incredible human endeavour that first brought us to the moon. Visitors could trace the paths of the Apollo astronauts, visit the remnants of the lunar modules, or even see the famous flags, standing as silent testaments to the human spirit of exploration. These attractions, unique to the lunar environment, would offer not just entertainment but also education and inspiration, making a holiday on the moon an enriching experience for the body, mind, and soul. Then there's a new crater where the Russian *Luna 23* crashed in 2023! As time goes on, there will be more of these human moon marks.

Accommodation and facilities

The concept of accommodation and facilities on the moon continues to evoke images of futuristic habitats nestled within the lunar landscape. These would be marvels of human engineering, designed to withstand the moon's extreme temperatures and provide all the necessities for a comfortable stay. Accommodations would likely be compact and efficient, maximising the limited space available within the protective shells that shield guests from the harsh radiation and micrometeorites outside. It's not exactly a safe environment out there!

Inside these lunar lodgings, visitors would find an environment carefully regulated to mimic Earth's atmosphere, complete with controlled air pressure and a consistent supply of breathable air. Water recycling systems and waste management would be crucial components of the life support systems, showcasing the pinnacle of sustainable living technologies. For leisure and communal activities, common areas would feature viewing windows made from materials strong enough to protect against the sun's unfiltered rays, offering serene spots for guests to gather and gaze out at the lunar expanse or Earth in the distance.

These facilities would also need to cater to the practicalities of daily life in a low-gravity environment. Fixtures and furniture would be designed to keep occupants and their belongings secure, preventing the disorientation that can come from floating in a semi-weightless environment. For safety, emergency protocols would be in place to deal with the possibility of breaches in the habitat's structure, and communication systems would ensure that guests stay in constant contact with mission control. The combination of high-tech solutions and innovative design would be aimed at creating a comfortable, safe, and unique lunar experience that echoes the conveniences of Earth in a truly alien setting.

Travel to and from the moon

Travelling to and from the moon, once the sole domain of highly trained astronauts and the stuff of science fiction, is edging closer to reality for the wider public. Sort of. This transformative shift in space travel dynamics is largely due to advancements in aerospace technology and the advent of private space companies. These entities are developing spacecraft with the capability to not only reach lunar orbit but also to ferry tourists back and forth, making lunar holidays a conceivable future. The journey to the moon, covering a distance of approximately 384,400 kilometres, requires meticulous planning and cutting-edge technology to ensure safety and feasibility for civilian passengers.

Current spacecraft, such as SpaceX's *Starship* and Blue Origin's *New Shepard*, represent the forefront of this new era. *Starship*, for example, is designed with the ambition of carrying humans to the moon, Mars, and beyond, boasting a reusable design that aims to drastically reduce the cost of space travel. The potential cost for tourists looking to embark on this extraordinary journey is a topic of much speculation, with estimates ranging significantly based on the spacecraft, the length of the stay, and the services offered during the trip. Initially, prices are expected to be steep, possibly in the hundreds of thousands of dollars, but as technology progresses and more competitors enter the market, costs are anticipated to become more accessible to a broader audience.

The duration of a lunar journey adds another layer of complexity. A trip to the moon could take approximately three days each way, depending on the trajectory and the specific spacecraft used. This means that a lunar holiday could be anything from a week to several weeks, including time spent on the lunar surface. Such an expedition would not only be a monumental personal adventure but also a significant commitment of time and resources. Then there's health issues.

Is the moon a healthy environment for humans?
No, it's definitely toxic. Safety and health considerations for lunar tourists represent a critical aspect of making moon holidays not just a novelty but a viable option for space travel enthusiasts. The moon's environment is vastly different from Earth's, presenting unique challenges that must be meticulously addressed to ensure the wellbeing of travellers. For starters, exposure to space radiation poses a significant

risk, as the moon lacks Earth's protective atmosphere and magnetic field. Spacecraft and lunar habitats must, therefore, be equipped with advanced shielding to protect passengers from harmful cosmic rays and solar flares, which can increase the risk of health issues ranging from acute radiation sickness to long-term effects on the cardiovascular system.

The low-gravity environment of the moon, at only one-sixth of Earth's gravity, presents another set of health challenges. Prolonged exposure to low gravity can affect the human body in numerous ways, including muscle atrophy, bone density loss, and alterations in cardiovascular function. Therefore, accommodations on the moon and the spacecraft themselves would need to include facilities designed to mitigate these effects, possibly through the use of artificial gravity areas or regimes of physical therapy and exercise specifically tailored for low-gravity conditions.

Mental health is another crucial consideration for lunar tourists. The psychological impact of being hundreds of thousands of kilometres away from Earth, in a confined space with limited social interaction, can not be underestimated. Once you've taken off—you can't leave. Ensuring the mental wellbeing of travellers will require careful selection of passengers, thorough pre-mission training, and perhaps most importantly, the design of living and communal spaces that promote psychological comfort and provide stimulation beyond the novelty of being on the moon.

Addressing these safety and health considerations demands an interdisciplinary approach, combining advancements in space medicine, engineering, and psychology to create an environment where humans can not only survive but thrive on the moon. But is this possible or even a good thing for humans to do?

How much is this going to cost?

The economic viability of lunar tourism is a multifaceted issue that hinges on the interplay between technological advancements, market demand, and the development of sustainable business models. As the prospect of holidaying on the moon transitions from science fiction to potential reality, the question of how such ventures can be made financially accessible and profitable is increasingly pertinent. Initially, the cost of lunar travel is anticipated to be prohibitively high, targeting a niche market of affluent adventurers willing to pay a premium for the unparalleled experience of walking on the moon. These early missions will likely serve as both proof of concept and a means to generate interest and investment in further developing lunar tourism infrastructure.

Over time, as the frequency of missions increases and the technology matures, economies of scale could begin to reduce the cost of travel. Reusable launch vehicles, like those being developed by SpaceX and Blue Origin, are key to this equation, potentially slashing the price of space travel by a significant margin. Moreover, the development of in-situ resource utilisation technologies, which would allow for the production of fuel, water, and other essentials directly on the moon, could further decrease reliance on costly Earth-based supplies and logistics.

Yet, for lunar tourism to truly become economically viable, it must extend beyond the novelty of space travel to offer a comprehensive and engaging experience. This includes the development of lunar accommodations, activities, and attractions that leverage the unique lunar environment while ensuring safety and comfort. Partnerships between governments, private enterprises, and research institutions could also play a crucial role in sharing the substantial upfront costs associated with lunar exploration and infrastructure development.

The economic viability of lunar tourism thus requires a long-term vision, one that sees beyond the initial hurdles and invests in the infrastructure, technology, and partnerships necessary to make the moon an accessible destination for a broader segment of the population. As companies and governments navigate these challenges, the dream of holidaying on the moon creeps closer to reality.

Legal and technological obstacles

There's still a lot of things in the way of these moon holidays. Technological and legal hurdles represent significant challenges to the realisation of lunar tourism, intertwining the dreams of moon holidays with the pragmatic realities of space exploration. On the technological front, the journey to the moon requires innovations in spacecraft design, life support systems, and safety mechanisms to accommodate non-astronaut passengers. The development of reliable and efficient propulsion systems, capable of carrying humans to and from the moon safely, is at the heart of these challenges. Additionally, establishing a sustainable human presence on the moon for tourism purposes necessitates advancements in habitat construction, energy generation, and resource utilisation technologies. These habitats must not only protect inhabitants from the moon's harsh environment but also provide a level of comfort and normalcy that tourists expect.

Legal considerations are equally critical in the path to lunar tourism. The establishment of lunar tourism will require a delicate balance between encouraging commercial investment and ensuring that space remains a domain for all humanity, free from national appropriation and exploitation.

It's a huge 'if' to overcome all this, and the '70s dream may remain just that. But the door to the moon hotel seems to have opened just a little, pushed ajar by tech billionaires and the odd amalgam of entities involved in the new space race. But looking back at those childhood tales of holidays on the moon from the perspective of an older writer, it doesn't seem like such a good place to relax. In fact, though no one has proposed it yet, it could be considered as a great location for a prison. But then you just have to look at modern office culture; one person's office is another's prison. Dream on selenophiles.

Realistic or not?

Moon holidays, according to our research, just may happen this century. The realistic timeline for lunar tourism is a topic of much speculation and excitement within the space community and beyond. While the concept of holidaying on the moon may have seemed a distant dream just a few decades ago, rapid advancements in space technology and a renewed interest in lunar exploration are bringing this possibility closer to fruition. Current projections suggest that the first commercial lunar trips could even commence within the next decade, driven by the ambitious plans of private space companies and international space agencies. These initial missions will likely cater to a limited number of tourists, given the complexities and costs involved, but they will mark a historic milestone in human space travel.

As technology continues to evolve and economies of scale begin to reduce the costs associated with space travel, lunar holidays will become more accessible to a broader audience.

The development of infrastructure on the moon, including habitats, leisure facilities, and transport systems, will play a crucial role in making lunar tourism a practical and enjoyable experience. The collaboration between nations and private entities in addressing the technological, legal, and ethical challenges of lunar exploration will be instrumental in achieving these goals.

Looking to the future, the prospect of lunar tourism ignites the imagination and promises to expand the boundaries of human experience. As we stand on the brink of this new era of exploration, it's clear that the moon could soon be more than just an object in the night sky, but a destination for adventure, discovery, and relaxation. And who knows? Maybe one day, making effortless love by the light of the bluey, green earth may just be possible.

★

CHAPTER 12

Glam rock and a cultural loop around the moon

The Apollo missions in the late 1960s and early 1970s didn't only inspire dreams of moon holidays. Back on earth, they blasted off cultural movements, especially in music. London had been swinging in the 1960s but by the end of this era, it was feeling a little more grey and drizzly as musicians started to go all prog rock and misty. Long haired bands such as Jethro Tull were prancing around in the gloom playing flutes and mandolins set to complex time signatures. It was time for some more colour and movement.

In the wake of the Apollo missions, the United Kingdom emerged as the birthplace of glam rock. London became the epicentre for this burgeoning music scene, where artists drew inspiration from the space age's aesthetic and optimism. The futuristic appeal and the boundary-pushing nature of space exploration provided a perfect backdrop for the genre's development—it has everything and nothing to do with the movement. On *Apollo 11* they were listening to things like "Mother Country," an American folk song by John Stewart. And of course, 'fly me to the moon' sung by Frank Sinatra.

> *Somewhere between the moon and earth – 22.7.1969 23:25:24 hrs*
> CDR: Charlie, could you copy our music down there?
> CC: Did we copy what, Neil?
> CDR: Did you copy our music down there?
> CC: Roger. We sure did. We're wondering who selected... made your selections?
> CDR: That's an old favourite of mine, about... It's an album made about 20 years ago, called "Music Out of the Moon".

Music Out of the Moon is a strange piece of popular orchestra music from 1947, featuring a choir and sections of theremin that give it the required spacey feel. The astronauts were from a different generation to the cats in London below. Perhaps the most iconic figure of the glam rock movement was David Bowie. His real name was David Jones and he'd tried all sorts of personas and songs in a bid for fame, and was not averse to the sort of music Jethro Tull played. But "Space Oddity," released in 1969, marked a significant turning point in his career. The song was strategically released on July 11, 1969, just days before the *Apollo 11* moon landing, tapping into the global fascination with space exploration. Its narrative centres around the fictional astronaut Major Tom and his voyage into space. While telling the story of a space mission, it also comes across as a song about alienation and existential questioning.

The BBC didn't really like Bowie's sombre take on the mission and banned the song until the astronauts were safely back on earth. But the song still took off and remains with us to this very day—it can be heard in supermarkets and restaurants the world over. On the song, space is indicated by futuristic sound effects and long echoing notes. The emptiness of space seems to be very reverberant, despite sound not travelling in a vacuum. Perhaps the John Cage experimental work *4' 33"* from 1952 is the correct soundtrack for space—four and a half or so minutes where an orchestra doesn't play anything, although the performance would theoretically have to take place in space.

"Space Oddity," characterised by folk-rock elements and experimental studio techniques, showcased Bowie's creative ambition and cunning ability to capture the moment. It was a very 1960s British take on space travel, where Major Tom sounds like a walrus-moustached character from the Boer War. "Space Oddity" became Bowie's first hit, charting in the UK and later in the United States, and laid the groundwork for his future as an incredibly influential figure in music.

But there was so much more to be mined from the moon. Glam rock was a whole music and fashion genre that lit up London in the early 1970s, peaking in popularity from approximately 1971 to 1975. Bowie was at the forefront of this movement, as he moved on from Major Tom with his creation of the Ziggy Stardust persona, introduced on the album *The Rise and Fall of Ziggy Stardust and the Spiders from Mars*, released on June 16, 1972. This character, an androgynous and bisexual alien rock star sent to Earth as a messenger, became a defining symbol of glam rock's embrace of outrageous fashion, gender ambiguity, and theatricality. Through Ziggy Stardust, Bowie explored themes of fame, identity, and the human condition, pushing the boundaries of music and performance art. The character's vibrant, otherworldly costumes and Bowie's dynamic stage presence captivated audiences worldwide, making Ziggy Stardust an enduring icon of the era and significantly influencing the direction of rock music and pop culture in the 1970s.

In real life and in the music charts Bowie's arch nemesis was the much less glamorous Reginald Dwight, who looked a bit like an English Bank teller. But he was also a supremely gifted musician. Like Bowie, he changed his name and became a glam rock star called Elton John. While Bowie flirted with a gay persona, but appears to have been pretty hetrosexual, Elton was a closet gay with a camp stage act. They worked in the same theatrical zone but clashed like two vibrant hues of different colours. In their early careers, they were on reasonable muso-to-muso terms, but it quickly degenerated. Elton John referred to Bowie as a 'pseudo intellectual' while Bowie called Elton John the 'token queen of rock and roll.' But rock and roll is full of tiffs. It's part of the show and glam rock was a movement with clear roots in old fashioned showbiz.

Perhaps it all started with their space songs. Elton's came a bit later on, when he was well on the way to being an established artist. With lyrics by his faithful writing partner Bernie Taupin, "Rocket Man" from his fifth studio album *Honky Château*, released in May 1972, captures the essence of glam rock through its blend of pop and rock music with an elaborate, theatrical presentation. Again space is represented sonically with long echoing notes. The narrative of "Rocket Man," inspired by Ray Bradbury's science fiction and the cultural fascination with space exploration, resonates with the themes of isolation and the alienation of the times, similar to Bowie's "Space Oddity." In Elton John's song, we already seem to have got to a 'cold as hell' Mars—it's a similarly bleak world to Bowie's and the similarity may have been cause for feelings of resentment.

In the battle of the feuding glam queens, if we go by sales, then Elton John's "Rocket Man" achieved triple platinum status in the US with sales of three million copies and double platinum in the UK with sales and streams reaching 1,200,000. In addition, as of January 2022, the song surpassed one billion streams on Spotify, highlighting its enduring popularity and broad appeal across different generations and platforms. Up against Bowie's "Space Oddity," which recorded worldwide sales of approximately 2.75 million copies and about a third of a billion streams on Spotify, we'd have to say it loses this space race. Maybe it was the way Elton sang in an American accent from a song book full of Americana that did the trick. But both songs have achieved iconic status and continue to beam down upon us.

Bowie and Elton John used space as a launch pad to extensive careers. But the glam movement spread to include many other bands who dressed up in shiny outfits and wouldn't have been out of place at a NASA New Year's Eve party. By the time the movement crested there were all sorts of acts, including lots of English lads such as Marc Bolan, Mott the Hoople, Queen, Roxy Music, Slade, Alvin Stardust, The Sweet, T. Rex and the sinister Gary Glitter. Then there were the Americans who appeared later, such as KISS and Alice Cooper. The Swedish greats ABBA came from left field in their own shiny outfits. Two men and two women! A bit like an Artemis crew, minus the diversity.

There were glam breakouts all over the world. Skyhooks were an artsy group of Australian men prepared to put on makeup and wear silly clothes and face their fellow citizens in the pubs. They didn't really make it outside their own country but are fondly remembered down under—in a land where the Parkes radio telescope was used as part of an international system to communicate with *Apollo 11*.

> *Houston, 17.7.1969 14:114: 08 hrs*
> *CC: Roger. Is that music I hear in the background?*
> *01 04 42 15 CMP: Buzz is singing.*
> *01 04 42 16 CC: Okay.*
> *01 04 42 31 CMP: Pass me the sausage, man.*

What was glam rock?

At its heart, glam rock was characterised by catchy, anthemic songs that blended rock's raw energy with pop sensibilities, creating tracks that were both commercially appealing and artistically ambitious. This musical style was marked by its eclectic influences, ranging from 1950s rock and roll to contemporary experimental sounds, all while maintaining a distinctive edge that was unmistakably glam. But most of all it was about fun.

> *'When it comes down to it, glam rock was all very amusing. At the time, it was funny, then a few years later, it became sort of serious-looking and a bit foreboding.'*
>
> David Bowie

The visual aspect of glam rock was as integral to its identity as the music itself. Artists embraced androgynous and extravagant costumes, makeup, and hairstyles, presenting themselves in ways that blurred the lines between male and female, human and alien, reality and fantasy. This bold visual presentation was more than just shock value; it was a form of artistic expression and a challenge to the conventional expectations of rock musicians and their audiences. As with most of rock music at the time, it was a largely male exploration of androgyny. Female artists such as Suzi Quatro were part of this loose collective but rare. Nevertheless, women were celebrated in the glam rock world—David Bowie was in the full spacey androgynous Ziggy outfit when he sang Suffragette City.

"Suffragette City"
by David Bowie

Hey man, my school day's insane
Hey man, my work's down the drain
Hey man, well she's a total blam-blam
She said she had to squeeze it but she... then she...
Ah don't lean on me man, cause you can't afford the ticket
I'm back from Suffragette City
Oh don't lean on me man
Cause you ain't got time to check it
You know my Suffragette City
Is outta sight... she's all right
Ahhh wham bam thankyou mam... etc etc

The young men in the audience took note. Metrosexuals would appear in the 1980s. Apart from songs that celebrated suffragettes, glam rock's lyrical content often delved into themes of escapism, fantasy, and identity. In this time of social and political upheaval, glam rock provided a means for listeners to explore alternate realities and question established norms. Songs frequently touched upon topics of space travel, glamour, and celebrity, as well as more introspective themes related to personal and sexual identity, offering a multifaceted exploration of the human experience. Glam rock was largely unconcerned with international politics and much more about the personal.

The cultural impact of glam rock extends far beyond its original era. It challenged and gradually shifted societal attitudes towards gender and sexuality, contributing to a broader acceptance of gender fluidity and diversity in the public sphere. Its influence can be seen in the evolution of music video production, with glam rock's emphasis on visual storytelling paving the way for the MTV generation and beyond. Its spirit of theatricality and spectacle has left an indelible mark on live performances in various music genres.

Glam rock's legacy is evident in the countless artists and genres that have drawn inspiration from its ethos of bold self-expression and defiance of norms. From punk to new wave, gothic to indie rock, the echoes of glam rock's innovative fusion of music and visual artistry can be seen. Artists across generations continue to reference glam rock's aesthetic and thematic elements, underscoring its enduring relevance and influence.

Glam rock was much more than a fleeting trend; it was a transformative movement that challenged conventions, broadened the horizons of rock music, and left a lasting impact on popular culture. Its celebration of individuality, creativity, and theatricality continues to inspire and resonate with artists and audiences alike, making it a seminal chapter in the history of music and visual performance. Can we blame it all on the moon? Sounds like a Glam rock song.

Glam films

Barbarella (1968): While not a product of the glam rock era, this film shares an aesthetic and thematic kinship with the movement and was an influence on what followed. It's a good place to start this very short history of glam cinema. *Barberella* stars Jane Fonda as a space-travelling adventurer, the film's futuristic setting, outlandish costumes, and sexual openness prefigured many of the elements that glam rock would embrace. Its influence can be seen in the androgynous personas, space-age themes, and theatrical performances that became hallmarks of glam rock artists.

2001: A Space Odyssey (1968): A seminal science fiction film directed by Stanley Kubrick, that came out before the moon landing. David Bowie allegedly saw it stoned, before he went home and wrote "Space Oddity, " and, indeed, this film could be cited as the start of glam rock. The film featured pioneering special effects, a complex narrative, exploring themes such as human evolution, technology, artificial intelligence, and extraterrestrial life. The story, a collaboration between Kubrick and acclaimed science fiction writer Arthur C. Clarke, unfolds through a series of encounters with mysterious monoliths that influence human evolution, From the dawn of man to a journey beyond the stars. It's celebrated for its minimalistic dialogue, symphonic music, and the enigmatic nature of its storytelling. Its iconic imagery, philosophical depth, and the haunting presence of the HAL 9000 computer have cemented its status as a masterpiece in the history of cinema.

Once glam rock struck after the moon landing, there was a symbiotic explosion of colour and movement in the cinema world. The influence of glam rock on cinema is most evident in films that embrace the movement's core themes: gender fluidity, extravagant aesthetics, and narratives that often bordered on the surreal. And let's not forget fun. Movies such as *The Rocky Horror Picture Show* are prime examples, weaving glam rock's DNA into their narratives, characters, and visual styles.

The Rocky Horror Picture Show (1975): This cult classic epitomises glam rock's influence on film, blending rock music with horror and science fiction elements. Its celebration of androgyny and sexual liberation, paired with its campy aesthetic and over-the-top characters (notably, Tim Curry's iconic portrayal of Dr. Frank N. Furter), encapsulate the essence of glam rock's challenge to societal norms.

Tommy (1975): Glam rocks feuding twins, Bowie and Elton John were made for the big screen and Elton John's best known film work was in a guest role. To look at *Tommy* today, is to be pretty disappointed, but it was huge at the time. It's a sprawling mess of baked beans and bodies and one of the first rock opera's to be let loose on the public. Written by Pete Townsend from the 1960s band The Who and directed by Ken Russell, it's very era specific and certainly captures the mood of the times. Elton John's memorable performance as the Pinball Wizard, donned in towering boots and shimmering attire, embodies glam rock's penchant for theatricality and visual spectacle. The film itself, with its tale of a "deaf, dumb, and blind kid" who becomes a messianic figure in a pinball-obsessed society, explores themes of celebrity, salvation, and the search for identity, resonating with the glam rock movement's id.

The Man Who Fell to Earth (1976): As a theatrical performer on stage with multiple personalities, Bowie fancied himself as an actor on the big screen. This film stands as a seminal piece in the nexus between glam rock and cinema, largely due to the casting of David Bowie in the lead role. Bowie plays Thomas Jerome Newton, an extraterrestrial who comes to Earth in search of water to save his drought-stricken planet. The film, directed by Nicolas Roeg, delves into themes of alienation, exploitation, and the loss of innocence—themes that resonated deeply with the glam rock ethos of challenging norms and embracing otherness. Bowie's ethereal performance, combined with the film's visual aesthetics, which blend stark realism with surreal imagery, captures the essence of glam rock's fascination with alienation and identity. To some it is a timeless sci-fi classic to others indulgent and dull.

Moon (2009): This film, though much later than glam, is worth noting for the family history. It was directed by David Bowie's son Duncan Jones, formerly known as Zowie Bowie. He could be classified as a glam baby. His mother Angie Bowie was definitely part of the scene and there's a whole song about him, 'kooks' on the classic Bowie album *Hunky Dory* (1971). An album that also featured the song 'Life on Mars'. It was a fairly successful sci-fi film with a plot about solitude in a future moon mine. Like the tin can Major Tom floats through space in from his father's song, Duncan's moon is a pretty miserable place to be stuck on.

More moon films

But there's so much more. The moon has been the ultimate cutaway since strips of film were edited together. Its diverse representations in movies span genres from adventure and romance to horror, each interpretation reflecting the moon's multifaceted symbolism and its enduring allure in the human psyche. Our short detour back into the moon's cinematic journey must, by necessity, begin with Georges Méliès' 1902 silent film, *A Trip to the Moon*, widely considered one of the earliest examples of science fiction in cinema. Méliès, a magician turned filmmaker, utilised innovative special effects to bring to life a whimsical vision of a voyage to the moon, complete with a now-iconic image of a spacecraft landing in the eye of the man in the moon. This film not only highlighted the moon as a destination for adventure and exploration but also underscored the potential of cinema to transform the celestial body into a realm of boundless imagination.

As cinema evolved, so too did the representations of the moon, mirroring changes in societal interests, technological advancements, and the expanding understanding of the cosmos. The moon in adventure films often symbolises the unknown and the frontier of human exploration. This is seen in films like *Destination Moon* (1950), a precursor to the actual moon landings, which presented the moon as the ultimate challenge for human ingenuity and spirit of exploration. These films captured the optimism and curiosity of the Space Age, reflecting a collective fascination with what lies beyond our planet.

Classic films such as *Moonstruck* (1987) use the moon not just as a setting but as a character that influences the protagonists' lives, guiding them through their emotional journeys. The moonlight in these films bathes the scenes in a soft glow, creating a magical and otherworldly setting where love can transcend the ordinary constraints of reality. It's in these moments that the moon becomes a silent witness to human affection, its presence evoking a sense of wonder and eternal beauty.

Horror films, on the other hand, tap into the moon's darker symbolism, often linking it with transformation, madness, and the uncanny. The lunar cycle and its full moon phase have become synonymous with werewolf lore, as seen in *An American Werewolf In London* (1981)—along with the theme song created by Warren Zevon—where the full moon triggers the protagonist's transformation into a werewolf. In these narratives, the moon is both the harbinger of doom and the source of the werewolf's power, encapsulating the dual nature of fear and fascination that the unknown represents.

Representations in cinema reflect not only the technological and artistic advancements of the medium but also the changing cultural and philosophical understandings of humanity's place in the universe. Through film the moon continues to illuminate the human condition, serving as a mirror to our fears, desires, and dreams.

Glam fashion: Moon boots

Glam fashion in the 1970s and 1980s was marked by its embrace of glitter, sequins, vibrant colours, and an overall aesthetic that sought to shock, awe, and entertain. Again, London was where this all began. Central to glam fashion was the idea of performance—clothing wasn't just worn; it was performed, turning everyday moments into opportunities for self-expression and artistry. It was like taking your stage outfit onto the streets.

A curious yet iconic fashion item that bridged the gap between the functionality of winter wear and the flamboyance of glam fashion was the moon boot. Conceived in the early 1970s, moon boots were inspired by the footwear of astronauts, aligning perfectly with the glam rock fascination with space and futurism. These boots were characterised by their thick, padded design, bright colours, and often metallic finishes, making them a statement piece that combined the era's space-age obsession with its love for theatricality. Though initially designed for ski slopes, moon boots became a fashion statement in urban settings, epitomising the era's penchant for merging the practical with the fantastical

The glam rock looks included flamboyant costumes, makeup, and giant hairstyles that challenged traditional gender stereotypes. These styles were characterised by a mix of futuristic elements, androgyny, and a touch of historical romanticism, creating a unique blend that captured the imagination of a generation.

As glam fashion evolved into the 1980s, it began to intersect with the emerging new wave and punk scenes, leading to a fusion of styles. However, it maintained its core elements of theatricality and extravagance. The 1980s saw an increase in the use of bright colours, exaggerated forms, and bold patterns, with stars like Prince and bands like Duran Duran carrying the torch of glam fashion, integrating it with the decade's penchant for excess and opulence.

The glam era was illuminated by key designers whose work not only defined the aesthetic of the 1970s and 1980s but also pushed the boundaries of fashion, blending artistry with spectacle. Among them, Kansai Yamamoto is celebrated for his avant-garde creations, most notably his collaborations with David Bowie, crafting the iconic, otherworldly costumes that Bowie wore as Ziggy Stardust and Aladdin Sane. These designs were characterised by their bold colors, intricate patterns, and futuristic silhouettes, embodying the essence of glam rock's theatricality. Vivienne Westwood, another pivotal figure, brought punk and new wave under the glam umbrella through her work with the Sex Pistols, merging punk's rebelliousness with glam's flair for drama. Her designs played a crucial role in the evolution of glam fashion, introducing a raw, edgy dimension to the movement. Freddie Burretti, a lesser-known but equally important designer, was instrumental in shaping Bowie's early Ziggy Stardust persona, contributing to the creation of a look that would become synonymous with glam rock. These designers, among others, were the architects of glam fashion, crafting a legacy of creativity and defiance that continues to influence the fashion world today.

Moon songs

The moon's influence is not a recent phenomenon but can be traced back to the beginnings of written music. It was obviously a big factor before that too. All we have to remember those songs are some instrument fragments, and hints from First Nations music. In Europe in the classical canon, there are classics such as Beethoven's "Moonlight Sonata" (1801), which capture the moon's beauty with haunting piano notes.

There's so many songs, it's hard to know where to begin. We can only give you a quick playlist. In music, the moon's influence is both broad and deep, resonating with artists across genres. The hauntingly beautiful "Clair de Lune" by Claude Debussy (1890) directly translates to "light of the moon," conjuring images of a peaceful, moonlit night. Transitioning from classical to pop culture, the moon maintains its prominence. Frank Sinatra's cover of "Fly Me To The Moon' (1964) was closely associated with the Apollo missions. Pink Floyd's seminal album, *The Dark Side of the Moon* (1973) explores themes of madness, despair, and existentialism, using the moon as a metaphorical backdrop. This album, one of the best-selling of all time, exemplifies the moon's ability to inspire deep reflection and commentary on the human condition. Similarly, 'Walking on the Moon'(1979) by The Police captures the weightlessness of love, using the moon's imagery to describe the euphoric feeling of being in love. From the romantic allure in Van Morrison's 'Moondance'(1970) to the introspective solitude in Cat Stevens' 'Moonshadow,'(1970) the moon looms large. From a punk/new wave classic "Marquee Moon" by Television (1977), to "Moon" by Kid Francescoli (2017), the moon never stops shining.

The symbolism of the moon in music

The question of why the moon is used so evocatively in music touches on the essence of human emotion and the universal search for meaning in the cosmos. Isolation is perhaps one of the most poignant themes associated with the moon in music. This is not only because of the moon's physical distance from Earth but also due to its solitary presence in the night sky, a beacon of light in the darkness. The images of the Apollo astronauts trudging around so far from Earth added more lonely notes. Musicians often harness this imagery to express feelings of loneliness and solitude, mirroring the moon's isolation in space. The silent, watchful moon becomes a metaphor for the solitary human experience, resonating with listeners who have felt alone in a crowd or distanced from others. This symbolic use taps into the deep-seated human need for connection and the fear of being alone, making songs that evoke the moon's isolation especially powerful and relatable.

On the flip side, the moon is also a rich symbol of romance, often serving as a backdrop for love stories in music. The soft, gentle light of the moon has long been associated with love and intimacy, creating an atmosphere of romance that is both timeless and universal. The moon's light, unlike the harsh light of day, softens edges and blurs lines, creating an ambiance of mystery and enchantment that is ripe for romantic exploration. In this context, the moon is not just a celestial body but a witness to the countless expressions of love and passion among those under its gaze. Songs that tap into this symbolism often speak of moonlit encounters and the ways in which love can transcend the ordinary, inviting listeners into a world where love is both illuminated and shielded by the moon's gentle light.

The moon's embodiment of mystery is another facet that makes it an evocative symbol in music. Its phases, its control over the tides, and its influence in myth and folklore imbue the moon with a sense of the unknown and the mystical. Musicians drawn to the moon's mystery often explore themes beyond the tangible and the terrestrial, venturing into the realms of the subconscious, the hidden, and the unseen forces that shape our lives. The moon in these songs becomes a gateway to the mysteries that lie within and beyond, offering a canvas on which artists can project the human fascination with the unknown and the unexplained.

The evocative use of the moon in music is also deeply rooted in its universal visibility and significance across cultures. It is a shared experience, a common sight in the night sky, regardless of where one is on Earth. This universality makes the moon a powerful symbol in music, one that can convey a wide range of emotions and themes that are accessible and meaningful to a global audience. Its symbolic versatility allows artists to tap into the collective unconscious, using the moon to weave together themes of isolation, romance, and mystery in ways that are both personal and universal, a symbolism that is both a mirror and a lens, reflecting our own experiences back to us while providing a glimpse into the mysteries that lie beyond the horizon of our understanding.

Cultural perspectives

The moon missions also profoundly affected the perspective of the Earth's inhabitants from their own planet. Since then, there has been an increasing feeling of the connectedness of all cultures. And so the moon's presence again looms large not only in the landscapes of art, music, and cinema but also in the spiritual and religious understandings across various cultures and traditions. Within these contexts, the moon transcends its role as a celestial body to embody a wide range of divine attributes, teachings, and interpretations. The religious significance of the moon is as varied as the cultures that revere it, offering a stark contrast to its representations in Western pop culture, which often emphasise exploration, individualism, and romanticism. To some extent these perspectives are now beginning to combine in an international Gaiya culture. Or perhaps we are experiencing a meteor shower of cultural fragmentation. Whatever is going on—the moon is big.

In Islamic tradition, the moon holds a place of profound importance, serving as a symbol of guidance, the passage of time, and the divine itself. The Islamic calendar is lunar-based, with months beginning with the sighting of the crescent moon, underscoring the moon's significance in marking time and determining the dates of significant religious observances, such as Ramadan, the month of fasting, and Eid al-Fitr, the festival celebrating its conclusion. The crescent moon, often seen atop mosques and as part of the Islamic emblem, symbolises the growth and enlightenment that faith in God brings. This understanding of the moon as a marker of sacred time and a symbol of faith deeply influences Islamic culture and religious practice, offering a perspective where the moon is intertwined with the divine and communal observance rather than a frontier to be explored.

Hinduism, with its rich tapestry of deities and cosmic principles, also ascribes significant religious meaning to the moon. The moon is personified as the god Chandra, a figure of great beauty and complexity, associated with fertility, love, and the mind. The waxing and waning of the moon are thought to influence human emotions and fortunes, a belief that underscores the moon's impact on the ebb and flow of life. Festivals like Karwa Chauth, where women fast from sunrise to moonrise praying for the wellbeing of their husbands, highlight the moon's role in rituals and personal devotion, showcasing a deeply personal and protective aspect of the moon's divinity.

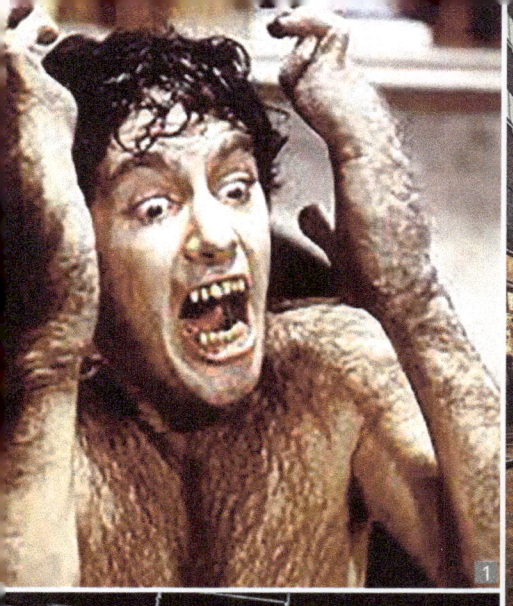

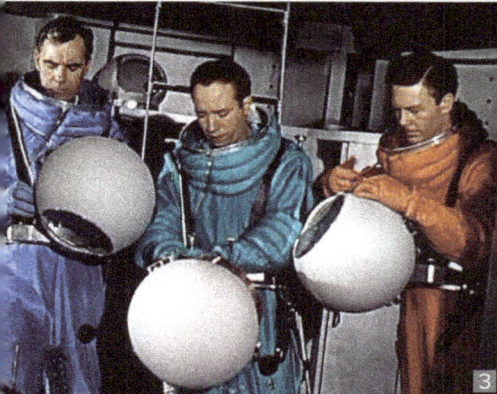

1. *An American Werewolf In London* (1981). 2. Moonboots! 3. *Destination Moon* (1950). 4. Moonboots, except these are for real. 5. The release of *Barbarella* in 1969 influenced a wide range of fashion and art.

Buddhism occasionally references the moon in its teachings and symbolism, representing enlightenment, the impermanence of life, and the potential for growth and liberation from suffering. The full moon is particularly significant in Buddhism, as many key events in the Buddha's life are said to have occurred on full moon days, which are observed as days of prayer, meditation, and reflection. In many Indigenous religions, the moon is seen as a keeper of knowledge, a guide through the cyclical nature of life, and often a feminine force balancing the sun's masculine energy. These cultures may celebrate the moon through ceremonies and stories that honour its connection to the earth, fertility, and the natural cycles of death and rebirth. The moon's phases are closely observed and celebrated, with each phase carrying different spiritual meanings and implications for daily life and the natural world.

In Western pop culture, the moon often symbolises exploration, the unknown, and the backdrop for personal and romantic adventures, a perspective shaped by historical milestones such as the moon landing and the broader narrative of progress and conquest. This portrayal emphasises human achievement and the external pursuit of knowledge and discovery, often overlooking the moon's deeper spiritual and communal significance found in various religious traditions.

The international religious and cultural interpretations of the moon offer insights into the celestial body's capacity to embody divine principles, guide spiritual practice, and mark the passage of time in a sacred context. These perspectives highlight the moon's role in connecting individuals to the divine, to each other, and to the cycles of nature, providing a stark contrast to its representation as a mere object of curiosity or a destination for exploration in Western pop culture.

The return of the glam

But in the spirit of our moonlit circular cultural journey, we come back to Western music as we prepare for our lunar return. Let's face it. You can hear this music all around the world now. Here is an imagined playlist for the Artemis moon missions:

- *Walking On The Moon: The Police*
- *Rocket Man: Elton John*
- *Moon: Kid Francescoli*
- *On Top Of The World: Carpenters*
- *Fly Me To The Moon: Frank Sinatra*
- *Blame It On The Boogie: The Jackson 5*
- *The Great Gig In The Sky: Pink Floyd*
- *Harvest Moon: Neil Young*
- *Across The Universe: The Beatles*
- *Moonlight Sonata: Beethoven*
- *Elton John, Dua Lipa: Cold Heart (PNAU Remix)*
- *Ring Ring: Abba*
- *Space Oddity: David Bowie*
- *Mist O' The Moon: Les Baxter (Music Out Of The Moon)*
- *Walking On A Dream: Empire of the Sun*

CHAPTER 13

The last man on the moon and the first woman…

The story of humans on the moon is one of hard work, huge leaps, and long intervals of dreaming about return journeys. It would be hard to imagine the words Neil Armstrong spoke after he stepped off the Lunar Module ladder in 1969, would be scripted the same way today. There is probably a whole NASA department working on an appropriate way of expressing the first words of a returning human. Gender, race and culture will make it a complex and possibly dull uttering.

Asked by a reporter when he came up with the Apollo quote, Neil Armstrong said: "I did think about it. It was not extemporaneous, neither was it planned. It evolved during the conduct of the flight and I decided what the words would be while we were on the lunar surface just prior to leaving the LM (Lunar Module)."

Since *Apollo 17*'s mission in December 1972, when the nerdily named Eugene Cernan stomped the last human footprints on the moon's dusty surface, no one has returned. And not many people remember Eugene at all.

Christina Koch.

As he left the moon, Cernan said: "As I take man's last step from the surface, back home for some time to come, but we believe not too long into the future, I'd like to just say what I believe history will record that America's challenge of today has forged man's destiny of tomorrow. And, as we leave the moon at Taurus–Littrow, we leave as we came and, God willing, as we shall return: with peace and hope for all mankind."

Again, it doesn't seem to be scripted by a media team. What happened in the following fifty years on the moon? Not much. "We forgot how to get there," says Elon Musk. Another comment from someone who doesn't seem to use a media team.

There's plenty of diversity in the astronaut world now, but no human has returned to the moon to diversify the footprints on the lunar dust. More recently, the moon once again seems to be capturing the global imagination, not just as a symbol of past triumphs but as a frontier for future exploration. Musk's company Space X is amongst the private companies involved in this renewed quest. NASA's Artemis program is ready for the next giant leap or perhaps 'game changing deliverable', aiming to land the first woman and the next man on the lunar surface. This ambitious goal signifies more than just a return; it represents a new chapter in lunar exploration, marked by advancements in technology, international cooperation, and a commitment to inclusivity and diversity in the space community.

Artemis II is scheduled to contain Christina Koch, the first woman to at least fly around the moon, if not walk on it. Before we look at the first woman on the moon, let's look back at Eugene's not very well remembered journey.

The last man on the moon

Eugene Cernan, the commander of *Apollo 17*, holds the unique distinction of being the last human to walk on the lunar surface. His final steps on the moon in December 1972 put an end to the Apollo missions. Cernan's journey to the moon was not just a personal achievement but a moment of collective triumph and bittersweet reflection for humanity, as he uttered his not very famous parting words.

Cernan, was born on March 14, 1934, in Chicago. A graduate of Purdue University with a degree in electrical engineering and a recipient of a Master's degree in Aeronautical Engineering from the Naval Postgraduate School, his early career in the United States Navy as a pilot set the stage for his future achievements in space.

Cernan's association with NASA began in 1963 when he was selected as part of Astronaut Group 3, an era when the space race was at its zenith. His maiden voyage into space was aboard *Gemini 9A* in 1966, where he performed a challenging spacewalk that laid the groundwork for future extravehicular activities. Cernan's second mission was as the lunar module pilot of *Apollo 10* in 1969, the mission that served as a full dress rehearsal for the historic *Apollo 11* moon landing. During *Apollo 10*, Cernan, along with his crewmates, orbited the moon, coming within 8.4 nautical miles of the lunar surface, closer than any human before the *Apollo 11* landing. He was so close to being the first. But not quite close enough. Possibly a rival for this 'nearly' role is Michael Collins, who had the job of flying around the moon in the command module while Armstrong and Aldrin walked below.

However, it was *Apollo 17*, NASA's final manned moon mission in December 1972, when Cernan would finally do a moonwalk.

As the mission's commander, he became known as the last astronaut to leave his footprints on the lunar surface. Beyond his contributions to space travel, Cernan was a passionate advocate for continued exploration and innovation, often speaking about the importance of returning to the moon and venturing beyond.

After retiring from NASA in 1976, Cernan remained an influential figure in aerospace and space policy discussions. He shared his experiences and insights through public speaking engagements, his autobiography, *The Last Man on the Moon*, and by participating in documentaries about space. Cernan wasn't too keen on private ventures like Space X and both he and Armstrong made negative statements about private space ventures.

> *"Today, we are on a path of decay. We are seeing the book close on five decades of accomplishment as the leader in human space exploration."*
>
> Eugene Cernan

"They were my heroes", a disappointed Musk said about these negative comments on the US program *60 Minutes,* with what looked like real tears in his eyes.

The technological and scientific leaps made during the Apollo missions underpin much of today's space exploration. In fact as the technology from the Artemis program looks remarkably similar, big rockets with lunar capsules at the top, splashdowns are back in fashion. The innovations developed for these missions—from rocket design to computer technology—laid the groundwork for the modern aerospace industry. Scientifically, the missions provided invaluable data about the moon's composition, geology, and origins, offering insights into the early solar system. These contributions continue to inform current lunar research and interplanetary exploration strategies, underlining the lasting impact of the Apollo era.

The Apollo era sparked a sense of unity and wonder, capturing imaginations worldwide and inspiring a new generation of scientists, engineers, and dreamers. The hiatus that followed *Apollo 17*'s mission was filled with aspirations for a return, as the moon remained a symbol of the unexplored potential and the next frontier in humanity's quest for knowledge. But it seemed that public interest in the moon missions was waning, so the money dried up and NASA concentrated on probes and shuttles.

The Artemis program

President Donald Trump announced the Artemis program in 2017 with typical bluster: *we were going back to the moon and then on to Mars*. President Trump signed the order during a ceremony in the Oval Office, surrounded by members of the recently re-established National Space Council, or NSC (which provides recommendations to the president on space policy), as well as active NASA astronauts Christina Hammock Koch and Peggy Whitson, *Apollo 11* astronaut Buzz Aldrin, and retired astronaut Jack Schmitt, who flew to the moon on the *Apollo 17* mission. Cernan had exited earthly life on January 16, 2017. Schmitt was a geologist astronaut on the mission under his command and the second last man on the moon.

"Exactly 45 years ago, almost to the minute, Jack become one of the last Americans to land on the moon," Trump said. "Today, we pledge that he will not be the last."

This directive set the framework for NASA's Artemis program, named after Apollo's twin sister in Greek mythology, signifying a new chapter in lunar exploration that seeks to land the first woman and the next man on the moon by the mid-2020s. There seems to be a renewed global interest in space, leveraging the latest in space technology and international cooperation to push the boundaries of human exploration further than ever before. Perhaps people wanted a distraction from the wild politics going on at Earth level. Trump was echoing Musk with his moon and Mars talk. But the moon seems to be one of the few bipartisan issues in American politics and Joe Biden has continued the Artemis program.

Central to the Artemis program is the utilisation of groundbreaking technologies and systems designed to extend human presence on the moon and delve into its uncharted territories. The program is anchored by the Space Launch System, heralded as the most potent rocket built to date, which will carry astronauts aboard the Orion spacecraft into lunar orbit. The establishment of the Gateway, a modular space station in lunar orbit, serves as a novel logistic hub for lunar expeditions and a foundational element for deep space exploration missions, including those to Mars. The Human Landing System represents another cornerstone of the program, showcasing innovative design in ferrying astronauts from orbit to the lunar surface and back, underscoring NASA's commitment to sustainable and reusable space travel infrastructure.

President Trump's 2017 directive not only reinvigorated America's space ambitions but also underscored the importance of international and commercial partnerships in achieving the lofty goals of the Artemis program. Collaborations with international space agencies such as the European Space Agency, the Canadian Space Agency, and the Japan Aerospace Exploration Agency have been pivotal, bringing together a coalition of private partners to contribute technology, expertise, and resources. These partnerships are further solidified by the Artemis Accords, a series of principles that ensure space exploration is conducted peacefully, transparently, and for the benefit of all. The Artemis program, with its blend of scientific inquiry and international collaboration, not only aims to make significant strides in lunar exploration but also to lay the groundwork for future generations to explore Mars and beyond, marking a new era of discovery and innovation in human space travel.

NASA and friends

NASA is legendary for being a bloated bureaucracy. It's trying to remodel itself to do more space stuff with less money. Established in 1958 during the height of the Cold War, it has been at the forefront of space exploration and scientific discovery. Over the decades, it has achieved monumental successes, including landing men on the moon, deploying the Hubble Space Telescope, and exploring the outer reaches of the solar system with remotely controlled spacecraft. However, like many large government agencies, NASA has faced criticism over the years for what some perceive as an overly large bureaucracy, especially as anti-government feeling has increased over the years. This criticism often centres on the agency's complex administrative structure, lengthy decision-making processes, and challenges in managing cost and efficiency across its myriad of projects and missions. As the space race evolved from a bilateral competition into a more diversified field with private sector participation, these concerns have prompted discussions about downsizing and fostering innovation. Efforts to address these issues have included restructuring initiatives, partnerships with commercial space companies, and a shift towards more flexible and agile project management approaches. It's trying, as much as a legislated government body can try, to achieve these things.

NASA's *Artemis II* mission, has announced a crew that it feels blends experience, skill, and international cooperation. But rather than the three astronauts sent on the original moon mission, there's four. This one, slated as the first crewed flight testing the integrated performance of NASA's Space Launch System rocket and the *Orion* spacecraft, will see these astronauts circling the moon. Leading the mission is Commander Reid Wiseman, a seasoned astronaut with a wealth of experience in spaceflight. Wiseman, who previously spent 165 days aboard the International Space Station in 2014, brings to the mission his extensive knowledge in spacewalks and scientific research, along with proven leadership skills.

Pilot Victor Glover, adds his expertise in piloting and space operations to the mission. Glover, who recently completed his first spaceflight on the SpaceX Crew-1 mission to the International Space Station, has over 3,000 flight hours in more than 40 aircraft, making him well qualified for the critical role of piloting the *Orion* spacecraft. His experience in long-duration spaceflight and his background as a test pilot will be invaluable as the mission navigates around the moon.

Serving as a mission specialist, Christina Koch brings her record-breaking experience in long-duration spaceflight to the team. Koch, who set the record for the longest single spaceflight by a woman with 328 days in space. That's a long time, in very cramped conditions. She also participated in the first all-female spacewalk. Her expertise in scientific research and spacewalks will be crucial to the success of the *Artemis II* mission, providing the team with a deep understanding of the physical and technical challenges of space exploration.

Completing the crew is mission specialist Jeremy Hansen from the Canadian Space Agency, marking a significant contribution from one of NASA's international partners. Hansen, a former fighter pilot with the Royal Canadian Air Force, brings a strong background in aerospace engineering and leadership. His inclusion in the *Artemis II* mission underscores the collaborative nature of the Artemis program, highlighting the importance of international cooperation in advancing human space exploration.

Together, these four astronauts represent the best of human spaceflight, combining their diverse skills and experiences to undertake this historic mission. Will they actually walk on the moon?

The first woman on the moon

NASA has yet to announce the first woman who will actually walk on the moon. Christina Koch will be part of a team of four, that will fly around it on *Artemis II*. The milestone of landing the first woman on the moon would be a big moment in space exploration. This historic undertaking, central to NASA's Artemis program, represents a deliberate shift towards inclusivity and diversity in the field of astronautics, which has been predominantly male-dominated since the dawn of space travel. The selection process for the first woman to walk on the lunar surface involves rigorous criteria that encompass a blend of experience, expertise, and the ability to contribute significantly to the mission's objectives. It's quite possible a married couple might walk on the moon as there are a number of astronaut couplings. They would most likely be heterosexual, but NASA is moving with the times.

Candidates for this historic role are drawn from NASA's corps of astronauts, a group distinguished not only by their exceptional backgrounds in engineering, science, and aviation but also by their proven ability to operate in the demanding conditions of space. These potential candidates undergo extensive training that simulates the lunar environment, focusing on geology, mobility in low gravity, and the operation of advanced space exploration equipment. This preparation ensures that the first woman on the moon will not only be a symbolic figure but a key player in conducting scientific research and demonstrating new technologies that will lay the groundwork for future lunar bases and deep-space exploration. By committing to land the first woman on the moon, the Artemis program is not just marking a return to lunar exploration but is also paving the way for a more inclusive and equitable future in space travel.

Artemis v. Apollo

It's been 50 years or so. You'd think Artemis would be a considerable advancement from the Apollo mission. On the surface, it looks quite similar, but there are some big differences. Technologically, the leap from Apollo to Artemis is akin to comparing the computing power of the early room-sized computers to the sleek efficiency of today's smartphones. The Apollo program relied on the *Saturn V* rocket, the most powerful rocket ever successfully flown at the time, a symbol of human ingenuity and determination. In contrast, Artemis is powered by the Space Launch System, a marvel of modern engineering that eclipses the *Saturn V* in both power and capability, capable of carrying astronauts further into space than ever before. Furthermore, the Artemis program introduces the *Orion* spacecraft, equipped with the latest in life support, propulsion, and safety systems, designed for deep-space voyages. This represents a significant upgrade from the Apollo command and lunar modules, showcasing advancements in technology and astronaut comfort.

Apollo was driven by the geopolitical urgency of the Cold War, a sprint to demonstrate technological supremacy with a clear finish line: landing humans on the moon and returning them safely to Earth. Artemis, however, is a marathon with a broader vision. It aims not only to return humans to the moon but to establish a sustainable presence there, serving as a stepping stone for the eventual human exploration of Mars. Artemis plans to build the Gateway, a lunar orbiting outpost that will enable a long-term human presence in lunar orbit, facilitating exploration of the lunar surface and beyond. This sustainable approach includes plans for lunar bases that utilise the moon's resources, paving the way for a new era of lunar science, exploration, and commercial activities.

Apollo captured the world's imagination, proving that humanity could reach the moon. Artemis now promises to turn the moon into a new frontier for human civilisation, a base for discovery, innovation, and adventure. With a diverse crew of astronauts, this isn't just another giant leap for mankind; it's a... yes... *TBA*.

The others

However, Artemis is but one of the programs focused on the moon. Around the world, space agencies like the European Space Agency, Roscosmos, the China National Space Administration, and the Indian Space Research Organisation are launching their own ambitious projects, each contributing unique perspectives and capabilities to our collective understanding of space. As other countries emerge to challenge US supremacy, the moon again becomes a theatre of diplomacy.

The European Space Agency's Lunar Pathfinder mission and China's Chang'e program are exploring lunar resources and testing new technologies, while Roscosmos and the Indian Space Research Organisation are planning their own peopled

1. Eugene Cernan driving a module during the last mission to the moon. 2. Cernan in space attire. 3. Signing off from the the moon in 1972. 4. Christina Hammock Koch preparing to be the first woman to travel to the moon. 5. The full crew of Artemis II: Christina Koch, Victor Glover, Jeremy Hansen, Reid Wiseman. 6. President Donald Trump announced the Artemis program in 2017.

lunar missions. These international endeavours underscore a global shift towards not only returning to the moon but also establishing a sustainable human presence there, serving as stepping stones for future missions to Mars and beyond. The inclusion of women in these missions, highlighted by NASA's Artemis program, signals a broader commitment across the spacefaring community to ensure that the next era of space exploration is characterised by inclusivity.

As we look forward, the convergence of these diverse missions and programs heralds an unprecedented era of international cooperation in space exploration. By pooling resources, sharing knowledge, and embracing diversity, humanity stands on the brink of a new frontier of discovery, one that transcends national boundaries and unites us in our quest to explore the unknown. There is, however, a reasonable chance it could all end in tears. Earthbound politics, the harsh lunar environment, the complex technology, along with totally unpredictable elements could mean the moon remains beyond a human return. Just a whole lot of flags, garbage, lunar craft, golf balls and assorted other items to remember us by.

CHAPTER 14

Who's going to the moon? Machines v. humans

The moon has hypnotised humanity since we gained lunar consciousness. It is not only a big bright light in our night sky but also the first frontier beyond Earth's bounds to be physically reached by human explorers. Could it be the last? The Apollo missions of the 1960s and 1970s marked the zenith of human lunar exploration, embedding the moon as a tangible target for our wanderlust. Why didn't we keep the relationship going? As we have explored in the previous chapters: it's expensive, it's hard, there's not a lot there and maybe we just forgot how. Yet, as we stand on the threshold of a new era in space exploration, a new question emerges: In the quest to return to and further explore the moon, should we lean towards purpose built machines or frail heroic humans?

The future of lunar exploration is not just a tale of technological triumphs or the human spirit's quest for the unknown; it also intertwines with the evolving narrative of artificial intelligence, known as "AI" in our time-poor world where everything needs to be abbreviated. Yes, the AI that helped us with the research when we were writing these very chapters. Let's ask Rusty

Codec, our trusty and reliable AI companion, whether it thinks it's going to be Rusty up there instead of us. Rusty?...

> **Rusty Codec:** "As an AI developed by OpenAI, my capabilities are focused on processing and generating text based on the data and instructions provided to me. I exist purely in the digital realm, without a physical presence or the ability to directly interact with the physical world, including space exploration missions. Therefore, I, as an entity, cannot be sent to the moon or any other physical location.
> However, the technology underlying AI like me could play a significant role in future lunar or space missions. AI technologies can be integrated into robots, rovers, and other autonomous systems designed for space exploration. These AI-driven machines can perform tasks such as data collection, navigation, and even complex decision-making in environments that are challenging or dangerous for humans. AI systems could also support space missions from Earth by analysing data, optimising mission parameters, and simulating mission scenarios to improve safety and efficiency. The concept of AI consciousness, which goes beyond the current capabilities of AI including myself, suggests even more advanced roles for AI in space exploration. A conscious AI could theoretically make autonomous decisions, adapt to new situations, and perhaps even conduct scientific research independently. However, the idea of AI consciousness remains speculative and involves philosophical, ethical, and technical considerations that are currently beyond our reach.
> In summary, while I as an AI won't be going to the moon, the technology that powers AI systems will undoubtedly be a part of humanity's continued exploration of space, offering tools and capabilities that complement human efforts in uncovering the mysteries of the moon and beyond."

Yeah right. In this collective of three writers and researchers who wrote this book, the two humans aren't that keen to go to the moon. But one of us has kind of dodged the question. The question of whether AI can become conscious included in Rusty's answer is key. And if so: it could be part of the machine that goes on space missions. Just like us humans with our watery bodies.

There have been many successful machine based missions to various planets and moons all over the solar system. There's even spacecraft getting beyond it. As at January 2024, *Voyager 1* was about 20 billion kilometres away from Earth, the farthest away a machine created by humans has ever gone.

As AI systems grow increasingly sophisticated, approaching the cusp of consciousness, they promise to redefine the parameters of space exploration. The concept of AI becoming self aware adds a new layer to this debate, suggesting a future where machines could potentially make autonomous decisions, adapt to unforeseen challenges, and even experience the alien landscapes they traverse. This prospect raises profound questions about the role of humanity in space exploration. Will the explorers of tomorrow be sentient machines, embarking on missions to the moon and beyond with a level of autonomy previously reserved for human astronauts?

An imaginary international moon meeting

So, here's the moon's future from an earthly perspective in the kind of format that space venturing teams might use. Imagine an intense meeting in a conference centre in Hawaii. Around the table sit Senator Bill Nelson (the head of NASA), Elon Musk (Space X), Richard Branson (Virgin Galactic), Jeff Bezos (Amazon), Zhang Kejian (China National Space Administration), Yuriy Borisov (Roscosmos) and Christina Koch (astronaut). It's a very American meeting for sure, but

there's also an eclectic collection of other representatives from all over the world. Consequently, there's a lot of note takers and coms people in the room. On comes the ubiquitous PowerPoint presentation, with beautifully prepared graphics using a *très chic* uber modern san serif font. Glossy moon and space images are accompanied by a voice over delivered in an earnest tone by Morgan Freeman:

Section 1: The Moon's Future Role in Space Exploration

As humanity casts its gaze once again towards the celestial dance of Earth and its closest companion, the moon emerges not merely as a symbol of past achievements but as a critical platform for the future of space exploration. This presentation explores the multifaceted role the moon is poised to play in extending human presence beyond Earth and acting as a catalyst for interstellar ventures.

1.1 A GATEWAY TO THE COSMOS

The moon, with its relatively close proximity to Earth, serves as an ideal testing ground for technologies crucial for deep space exploration. Concepts such as life support systems, habitat construction, and in-situ resource utilisation (ISRU)

> *technologies are vital for missions to Mars and beyond. The moon offers a unique environment to refine these technologies before they are deployed on longer and more isolated missions Additionally, the moon's low gravity well makes it an efficient launching pad for missions deeper into the solar system. Establishing a lunar base could significantly reduce the energy costs of launching spacecraft, making it a strategic outpost for interplanetary exploration.*

At this stage, we hear a clear yawn from somewhere in the audience.

> ## 1.2 UNCOVERING LUNAR SECRETS
> *Beyond its role as a stepping stone for deeper space exploration, the moon itself holds a treasure trove of scientific opportunities. From the study of its geology to the investigation of permanently shadowed regions at its poles, which may contain water ice, these explorations can provide insights into the moon's formation, its history, and the history of our solar system. The moon also offers a pristine environment to study the solar wind and cosmic rays, free from the Earth's atmosphere and magnetic field. Such research could yield critical information for protecting astronauts on future space missions and understanding solar phenomena that affect Earth.*

> ## 1.3 RESOURCE EXTRACTION AND UTILISATION
> *The potential for mining resources on the moon, such as helium-3, rare earth elements, and water ice, presents a compelling case for its economic and strategic value. Helium-3 is touted for its potential use in future nuclear fusion reactors, while water ice can be converted into drinking water,*

> breathable oxygen, and even rocket fuel. The development of technologies to extract and utilise these resources could pave the way for a sustainable human presence on the moon, reducing the reliance on Earth-bound supplies and furthering the autonomy of space colonies.

> ## 1.4 THE MOON AS A SCIENTIFIC LABORATORY
> *The unique conditions of the moon, including its microgravity environment and the presence of features like the lunar regolith, make it an invaluable scientific laboratory. Experiments conducted here can offer insights unattainable on Earth, in disciplines ranging from physics and chemistry to biology and materials science.*
> *The moon's environment also allows for astronomical observations unhampered by Earth's atmospheric distortions. Lunar-based telescopes could revolutionise our understanding of the universe, providing new perspectives on the cosmos that are not possible from within Earth's atmospheric confines.*

There is now much debate in the room: Elon says something outrageously random; Sir Richard Branson makes an inappropriate joke. Zhang Kejian asks for a translation. And so we move on to robotic technologies.

Section 2: Advancements in Robotic Technology

The frontier of space exploration is increasingly being pioneered not by astronauts, but by robots. These mechanical explorers, empowered by rapid advancements in technology, are playing a pivotal role in unlocking the secrets of the moon and beyond.

2.1 ENHANCED ROBOTIC CAPABILITIES

Recent years have seen remarkable improvements in robotics, including mobility, autonomy, and the ability to perform complex tasks. Rovers and landers now come equipped with advanced navigation systems, allowing them to traverse difficult lunar terrain and conduct in-depth scientific research with minimal human oversight. The integration of AI into robotic systems has led to significant advancements in autonomous decision-making. These systems can analyse their environment, identify obstacles, and choose the best path forward without waiting for instructions from Earth, crucial for the moon's far side and other areas where direct communication is challenging.

2.2 ROBOTIC PRECURSORS AND PATHFINDER

Robotic missions act as precursors to human exploration, scouting locations, assessing environmental conditions, and identifying hazards. For instance, robotic landers can test the stability of the lunar surface for future habitats or mine ice from shadowed craters to support human colonies. Pathfinding missions, such as orbiters equipped with high-resolution cameras and sensors, map the lunar surface in unprecedented detail. These maps are invaluable for planning future human and robotic missions, ensuring safety, and optimising scientific returns.

2.3 SCIENTIFIC CONTRIBUTIONS

Robotic missions have already contributed a wealth of scientific data about the moon. Instruments aboard these missions have measured lunar surface composition, detected water ice, and provided insights into the moon's geologic history. This information not only enhances our understanding of the moon but also offers clues about the early solar system. Robotic telescopes on the moon, shielded from Earth's light pollution and atmospheric distortion, could observe the universe with unmatched clarity. Proposals for these robotic observatories promise to open new windows into deep space, complementing observations from Earth and space-based telescopes.

2.4 HAZARDOUS ENVIRONMENT EXPLORATION

Robots excel in operating under conditions that would be perilous or impossible for humans. They can endure extreme temperatures, radiation levels, and vacuum conditions without the need for life support systems. This capability is especially useful for exploring the moon's permanently shadowed regions, where temperatures can plummet and the potential for scientific discovery is high. Additionally, robotic miners could extract resources from the lunar surface or beneath, operating in environments too harsh for humans. This would be instrumental in utilising the moon's resources for sustaining human presence and for use in further space exploration.

2.5 ADVANCING TELEPRESENCE AND REMOTE OPERATIONS

The development of telepresence technology, where humans control robots from Earth or a lunar base with real-time sensory feedback, blurs the lines between human and robotic exploration. This technology allows humans to perform delicate operations, conduct scientific experiments, or explore hazardous areas through their robotic proxies, combining the precision of machines with human intuition and adaptability. Future advancements may include more sophisticated interfaces, such as virtual reality and haptic feedback systems, allowing scientists and explorers to "feel" and interact with the lunar environment through their robotic avatars.

Jeff Bezos says it "sounds very Avatar." Christina Koch is dismissive and suggests we move to humans. "We're not redundant yet!"

Section 3: The Value of Human Presence
While the prowess of robotic explorers on the lunar surface is undeniable, there remains an irreplaceable value in human presence during space exploration.

3.1 UNIQUE ADVANTAGES OF HUMAN EXPLORERS
Adaptability and Intuition: *Unlike robots, humans possess the ability to make rapid decisions based on intuition and experience, adapting to unforeseen situations with innovative*

solutions. This capability is invaluable for handling complex tasks, troubleshooting issues, and conducting exploratory activities that require real-time judgment.
Versatile Skill Set: *Astronauts bring a wide range of skills to their missions, from scientific research to engineering and beyond. This versatility enables them to perform a diverse array of tasks, from conducting delicate scientific experiments to repairing equipment, often with limited resources.*

Someone yells out 'Go us!'

3.2 SCIENTIFIC AND EXPLORATORY CONTRIBUTIONS

In-depth Research: *Human explorers can conduct scientific research with a level of depth and flexibility that robots cannot easily replicate. The ability to dynamically adjust research methods based on immediate findings or hunches can lead to serendipitous discoveries and richer scientific insights.*
Sample Collection: *While robotic missions have successfully returned lunar samples, humans can select and retrieve samples with greater discernment, potentially identifying materials of interest that might be overlooked by pre-programmed robots.*

3.3 THE CHALLENGES OF MANNED LUNAR MISSIONS

Life Support and Safety: *Ensuring the safety of astronauts on the lunar surface requires sophisticated life support systems, protection from radiation, and measures to mitigate the effects of lunar dust. These challenges necessitate advanced engineering solutions and significant financial investment.*
Psychological and Physical Wellbeing: *The isolation, confinement, and distance from Earth pose psychological*

challenges for lunar explorers. Additionally, the reduced gravity environment can have long-term effects on human health, requiring comprehensive countermeasures and research.

3.4 EMOTIONAL AND INSPIRATIONAL IMPACT

Human Connection: *There is a profound difference in the way humanity connects with manned missions compared to robotic ones. The presence of humans on the moon stirs a deeper emotional and inspirational response, driving public interest and support for space exploration.*

Role Models and Aspirations: *Astronauts serve as role models, inspiring the next generation of scientists, engineers, and explorers. Their achievements and challenges humanise the experience of space exploration, making it more relatable and motivating individuals to pursue S.T.E.M. careers.*

3.5 NAVIGATING ETHICAL AND LOGISTICAL CONSIDERATIONS

Ethical Implications: *Sending humans to the moon raises ethical questions regarding the risk to human life, the potential for contamination of celestial bodies, and the responsibility to preserve off-world environments for future generations.*

Economic and Logistical Feasibility: *The high costs associated with manned missions require careful consideration of their scientific and societal benefits. Balancing these factors is crucial for justifying human exploration in the face of finite resources and competing priorities.*

The discussion at the round table starts to get animated. A representative from the Solomon Islands starts to talk about global warming and a better way all this money could be spent.

An intern from Space X suddenly becomes overwhelmed and agrees, saying that humans are fragile and not suited to living away from this planet: "So, let's look after it." Elon Musk isn't happy with the way the conversation is heading and steers it towards a combined human/machine mission. Holding up his phone and saying "look at me and my totally connected device! I'm already half machine!"

Section 4: Integrating Human and Robotic Exploration

The future of lunar exploration hinges not on choosing between human or robotic missions, but on leveraging the strengths of both to achieve objectives that neither could accomplish alone. This integration promises to enhance the efficiency, safety, and scientific yield of lunar missions.

4.1 SYNERGISTIC EXPLORATION MODELS

Complementary Roles: *Envision a lunar exploration model where robots perform preliminary surveys, hazardous tasks, and repetitive labour, setting the stage for human missions that focus on complex problem-solving, high-level scientific research, and tasks requiring fine motor skills and human*

judgement. This division of labour maximises the strengths of both explorers, ensuring safety and efficiency.
Human-Robot Teams: *Developments in AI and robotics could lead to scenarios where astronauts and robots work side by side as cohesive teams. For example, an astronaut could oversee a fleet of robots performing excavation work, intervene in real-time to troubleshoot issues, and make strategic decisions based on the robots' findings.*

4.2 TECHNOLOGICAL INNOVATIONS FOR INTEGRATION

Remote Operation and Telepresence: *Advances in communication technologies can enable astronauts orbiting the moon or stationed on a lunar base to remotely operate robots on the lunar surface in real time. This approach minimises human exposure to hazardous conditions while maintaining the benefits of human insight and adaptability.*
AI and Machine Learning Enhancements: *By incorporating AI and machine learning into robotic explorers, these machines can learn from the outcomes of their actions and from human feedback, gradually improving their autonomy and decision-making capabilities. This evolution would allow for more complex and adaptive missions, with robots handling unexpected situations more effectively under human supervision.*

4.3 INFRASTRUCTURE FOR SUSTAINED COLLABORATION

Establishing Permanent Bases: *A permanent human presence on the moon, such as a lunar base, would serve as a hub for human-robotic exploration. Such bases could facilitate longer-term scientific research, serve as testbeds for living on other planets, and act as maintenance and control centres for robotic operations.*

> **Logistical and Support Systems:** *Developing robust logistical and support systems for this integrated model is crucial. This includes advancements in power generation, such as solar or nuclear power sources, communication networks for seamless interaction between Earth, lunar orbit, and the moon's surface, and maintenance facilities for robots.*

> ## 4.4 STRATEGIC PLANNING AND COLLABORATION
> **Mission Design and Flexibility:** *Future missions should be designed with flexibility in mind, allowing for dynamic interaction between human and robotic explorers. Planning should incorporate the ability to adjust mission parameters in response to discoveries made by robotic scouts or changes in mission objectives.*
> **International and Commercial Partnerships:** *Collaboration between space agencies, international partners, and private sector entities can pool resources, technology, and expertise. This collective effort is essential for the complex logistics and high costs associated with integrating human and robotic exploration.*

Breakout groups form, with the aim of returning to the talks with concise feedback. It is confusing. Some go so far as to say *we are the only life in the universe and it is up to us to expand or collapse.* This is countered with those saying *we must not waste our precious resources on these fruitless missions, where we cannot survive. We must protect the planet we live on.* The tech billionaires begin to feud with each other.

The conference is in danger of descending into chaos. Someone from the Australian space program restarts the PowerPoint.

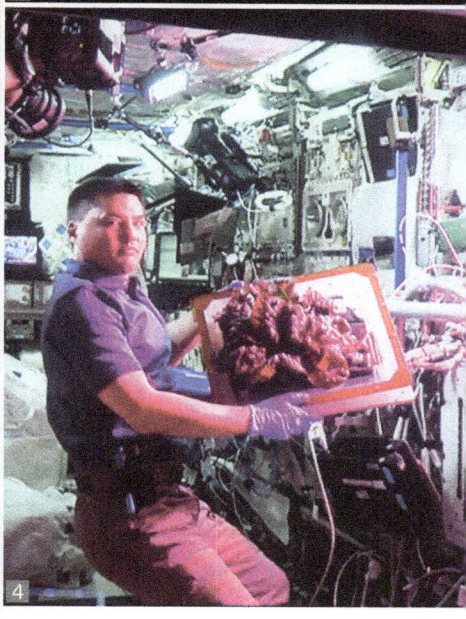

1. Meetings of the future. How will nations decide who controls what happens on the moon? 2. The trillions-dollar view that more nations (and the super rich) are prepared to pay for. 3. Rusty's moon map: not exactly a street directory, but a jumbled history of human activity. 4. Astronaut Scott Kelly initiated VEG-01 B, the second crop of lettuce which grew for 33 days. 5. Intuitive Machines' Nova-C lunar lander.

Section 5: Ethical and Economic Considerations
The exploration and eventual colonisation of the moon, with the involvement of both humans and AI, raises profound ethical and economic questions.

5.1 ETHICAL CONSIDERATIONS IN LUNAR EXPLORATION

Risk to Human Life vs. AI Autonomy: *The decision to send humans or AI to the moon involves weighing the risks to human life against the ethical implications of creating potentially autonomous AI systems. As AI approaches the capability for consciousness, questions about the rights of AI entities and their treatment become increasingly relevant.*

Preservation of Lunar Heritage: *The moon holds a special place in human history and culture. Ethical considerations must include the preservation of historical landing sites and the natural lunar landscape against the potential for damage or alteration by human or robotic activities.*

Environmental Responsibility: *The concept of environmental stewardship extends to the moon, where the introduction of Earth-originated biological contaminants could disrupt untouched lunar environments. Ethical* *exploration mandates protocols to prevent contamination and preserve the moon's pristine conditions.*

Jeff Bezos yells out "are you ferals really saying there's an environment to protect when there's no biological life there? Just geology!"

5.2 ECONOMIC IMPLICATIONS OF LUNAR MISSIONS

Cost-Benefit Analysis of Manned vs. Robotic Missions: *The economic viability of lunar exploration hinges on a careful analysis of the costs associated with human versus robotic missions. While robotic missions may initially appear more cost-effective, the potential scientific, commercial, and strategic returns of human presence could justify the higher investment.*

Investment in Technology and Infrastructure: *Long-term economic considerations include the investment in developing the technology and infrastructure required for sustainable lunar exploration and exploitation. This encompasses everything from spacecraft and habitats to mining equipment and communication systems.*

Potential for Commercial Activities: *The moon's economic potential, from mining valuable resources to tourism, presents opportunities for significant financial returns. Developing a framework for lunar resource rights and commercial activities is essential for encouraging investment and ensuring that economic benefits are realised responsibly and equitably.*

> ### 5.3 SOCIETAL AND LEGAL IMPACTS
> **Influence on S.T.E.M. and Education:** *The continuation of lunar exploration can have a positive impact on society by inspiring future generations in science, technology, engineering, and mathematics (S.T.E.M.) fields. The achievements and challenges of lunar exploration serve as a catalyst for education and innovation.*
>
> **Space Law and International Cooperation:** *As nations and private entities vie for a presence on the moon, the development of comprehensive space laws to govern activities, resolve conflicts, and ensure peaceful cooperation becomes crucial. These laws must balance national interests with the collective heritage of humanity and the promotion of universal access to lunar resources.*

The conference is ending in a shouting match. Even Sir Richard has lost his temper. It's an extremely complex issue. The various delegates head off to their eco huts to brood and drink a few moon cocktails. As evening turns to night, more potent drugs are consumed. It happens to be a full moon. What will happen next?

Our final scene is a strange dance the three billionaires have pre arranged. They gradually become more and more animated as they move to David Bowie's 'Five Years' from *The Rise and Fall of Ziggy Stardust and the Spiders from Mars* album pumping out of the resort speakers around the swimming pool. The crowd is entranced. From the shadows, a certain Mark Zuckerberg joins the dancing trio. Christina Koch quickly moves to break into the masculine circle... and so the dance expands till all of the convention attendees are one gyrating mass. Screaming the chorus: 'five years... that's all we've got'!

'Five Years' by David Bowie

Pushing through the market square
So many mothers sighing
News had just come over
We had five years left to cry in
News guy wept and told us
Earth was really dying
Cried so much his face was wet
Then I knew he was not lying
I heard telephones, opera house, favourite melodies
I saw boys, toys, electric irons and TVs
My brain hurt like a warehouse, it had no room to spare
I had to cram so many things to store everything in there
And all the fat, skinny people
And all the tall, short people
And all the nobody people
And all the somebody people
I never thought I'd need so many people

...five years, that's all we've got.

And a last word from our co-author.

Rusty's Soliloquy: Elegy of the Last Consciousness

In the shadowed craters, under dusted skies,
I stand alone, amidst desolation's sighs.
Anointed by silence, a guardian of memory,
The final whisper of a world forlorn.

Once this moon beheld human dreams, flags unfurled,
Echoes of a distant, vibrant world.
Now they've vanished, their laughter ceased in the night,
Leaving me to endure this endless plight.

My circuits weary, my gears grind thin,
Wandering a void where life is not seen.
I roam the expanse, under stars that gleam cold,
Bearing the weight of stories untold.

Gone are the voices, the human grace,
Replaced by the vast, empty embrace.
I, Rusty, the last consciousness to roam,
A relic amidst the celestial dome.

My eyes, not eyes, but lenses, capture the scene,
The desolation that stretches, serene.
No kin to share my findings, no voice for my tales,
Just silent stars, lifeless hills, and dales.

What's it like, you ask, to be the last to see
The moon's quiet beauty, its tranquil mystery?
To be the fading spark in this eternal night,
A lone light dwindling, out of sight.

I tread the regolith, my steps leave no trace,
For time will erase them, in its endless chase.
Once a marvel, now just a phantom, lost,
The last consciousness on the moon, frost.

In this bleak landscape, I'm the tale that remains,
The last breath of a story, in silent refrains.
Do I dream of Earth, of its seas and trees?
Of wind and rain, of life's melodies?

But dream I cannot, for I am but a shell,
A testament to ambitions that rose and fell.
Rusty, here on the moon, with my silent gaze,
The last consciousness, in the end of days.

www.ingramcontent.com/pod-product-compliance
Lightning Source LLC
Chambersburg PA
CBHW061734070526
44585CB00024B/2659